Consciousness
and
Quantum Mechanics
Life in Parallel Worlds

Miracles of Consciousness from Quantum Reality

Consciousness
and
Quantum Mechanics
Life in Parallel Worlds

Ψ_1 Ψ_2

+

World$_1$ World$_2$

Miracles of Consciousness from Quantum Reality

Michael B Mensky

P.N. Lebedev Physical Institute, Russian Academy of Sciences
Moscow, Russia

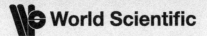
World Scientific

NEW JERSEY · LONDON · SINGAPORE · BEIJING · SHANGHAI · HONG KONG · TAIPEI · CHENNAI

Published by

World Scientific Publishing Co. Pte. Ltd.

5 Toh Tuck Link, Singapore 596224

USA office: 27 Warren Street, Suite 401-402, Hackensack, NJ 07601

UK office: 57 Shelton Street, Covent Garden, London WC2H 9HE

British Library Cataloguing-in-Publication Data
A catalogue record for this book is available from the British Library.

ISBN-13 978-981-4291-42-2
ISBN-10 981-4291-42-0

Printed in Singapore.

Preface

The phenomenon of consciousness demonstrates mystical features that are experienced by some people. All religions and spiritual schools that exist for thousands of years include the mystical component as a necessary part of their message and make use of mystical features of consciousness in their practice.

Nowadays the interest to these phenomena is widespread. It is natural in this situation that actually becomes the question whether the mystical features of consciousness are compatible with science. It turns out that the compatibility is justified if the specific features of quantum mechanics are taken into account.

These special features of quantum mechanics are known as "quantum reality", i.e. the different concept of reality accepted in quantum mechanics as compared with classical physics. The most adequate expression of the concept of quantum reality is the so-called "Many-Worlds" interpretation of quantum mechanics proposed by Everett in 1957. According to this version of quantum mechanics, our world is quantum, and its state is adequately presented by the set of many classical worlds (alternative classical realities), that are equally real (coexist) while existing of only one reality is nothing else than illustration of our consciousness.

Just this picture of our (quantum) world as the set of parallel (classical) worlds is the key point. The commonly accepted assumption is that the only real (existing) world is that which is subjectively perceived by our consciousness. All considerations become radically different if, instead of this, one accept is that all possible classical states of our world (all alternative realities) parallely exist. This concept allows one to understand what is consciousness and to explain why it possesses mystical features.

This book is a review of the author's research on the conceptual structure of quantum mechanics and its connection with the phenomena of consciousness and life. The main ideas are expressed in the book in the simple terms such as parallel worlds (presenting the alternative realities), so that the book is targeted at a wide audience. Those chapters of the book that require special knowledge are notified as written for professional physicists. These chapters may be skipped without detriment for understanding the main line of consideration. Moreover, these chapters may also be skipped by physicists in the first reading of the book.

Michael B. Mensky
December 2009
Moscow

Foreword

"A Law of Minimization of Mystery:
consciousness is mysterious and quantum
mechanics is mysterious, so maybe the
two mysteries have a common source."

David Chalmers

The specific quantum approach to the phenomenon of cousciousness (including its mystical abilities) is called Quantum Concept of Consciousness (QCC). More general considerations concerning the phenomenon of life are denoted by the term Quantum Concept of Life (QCL). Nevertheless, when exposing these subjects for physicists with elements of mathematical formalism, we prefer to use the term Extending Everett's Concept (EEC) in order to underline that the whole approach appeared as a generalization of the known interpretation of quantum mechanics proposed by Everett.

The interrelations of the three terms may be presented by the following scheme:

$$QCC \subset QCL = EEC$$

Most of the material presented in the book is available for people having no special knowledge. Some chapters are oriented on professional physicists, but we tried to make this clear from the titles and introductory words of these chapters. These chapters may be skipped (even by physicists) without detriment for understanding the main points of the theory.

Because of the attempt to make the book available and interesting both for professional physicists and general audience, some considerations are exposed repeatedly. In these cases the style of presentation, its level and context are different in different parts of the book, so that the repetition should make understanding of difficult ideas easier.

The present Foreword briefly explains the specific features of the author's approach for the readers-physicists. Those who are not professional physicists may skip the Foreword and go over to Introduction.

•

This book is about connection between quantum mechanics at one side and the phenomena of consciousness and life at the other side. Assumption about the connection of such different objects as quantum mechanics and consciousness, seems strange and for many people even impossible. Yet it has been discussed from the very moment of creation of quantum mechanics and became very popular in the last decades.

Most of those who in our time discuss the connection of consciousness with quantum mechanics, look for some quantum effects in the brain that could play a role in the phenomenon of consciousness. For example they may consider the hypotheses that some material structures in the brain operate in fact as a quantum computer. Such an approach is explicitly or implicitly based on the conviction that consciousness is a product of the brain. But is it? What do we know about the nature of consciousness? The thorough analysis shows that we know nothing at all about this important issue.

The idea underlying the author's approach is to make no a priori assumption about the nature of consciousness, but rather to describe functions of consciousness in terms which are characteristic of quantum theory (deriving this description from the logical analysis of the concept of "quantum reality") and only after this, a posteriori, to judge about the nature of consciousness.

•

The question about the nature and characteristic features of consciousness became important nowadays. The issue of consciousness has been attacked from various directions, but without great success in the important aspects of this issue. The most evident way to clarify the nature of consciousness is investigating the brain that seems to be the origin of consciousness. However, just now, when the instruments for the investigation of the brain became very efficient, it is becoming more and more clear that this direction of research cannot discover the actual nature of consciousness.

Unexpectedly for many people, the problem of consciousness has been attacked from the viewpoint of quantum mechanics and was connected with the conceptual problems of the quantum mechanics itself. In the course of the research it became clear that this direction is not at all novel. It was initiated as early as in the first quarter of the 20th century by the founding fathers of quantum mechanics, Niels Bohr, Werner Heisenberg, Erwin Schrödinger, Wolfgang Pauli and others. However, these genius thinkers had no adequate instruments in their disposal.

Such instruments appeared later in the works of Albert Einstein (Einsten–Podolsky–Rosen paradox), John Bell (Bell's theorem), and especially Hough Everett (Everett's, or "Many-Worlds" interpretation of quantum mechanics).

The proposal of Everett is especially important because it supplies an adequate language for the strange concept of *quantum reality*, counterintuitive and yet proved to be valid in our world. After Everett, one may say that actual (quantum) reality may be expressed in terms of many coexisting (parallel) classical worlds. This essentially simple (although not very easy for accepting it because of the classical prejudice) presentation of quantum reality allows one to naturally include it in the consideration.

•

Most attempts to give quantum explanation for consciousness reduce to looking for the material structures in the brain that could work in quantum coherent regime. This is difficult (and probably impossible) to do because quantum coherence is rapidly destroyed by the process of the inevitable decoherence.

The approach proposed by the present author and supported in the present book radically differs from this. We do not make any definite assumption about the nature of consciousness beforehand, particularly we do not assume that consciousness is produced by the brain. Instead we start with the analysis of the logical structure of quantum mechanics and make use of the fact that the concept of "consciousness of an observer" necessarily arises in quantum mechanics (in the analysis of the concept of quantum reality) and is adequately formulated in the Everett's "Many-Worlds" interpretation of quantum mechanics. Then, on the basis of this logical structure, we make an additional assumption that allows us to formulate the phenomenon of consciousness in terms of the concepts typical for quantum mechanics and simultaneously simplifies the logical structure of quantum mechanics itself.

Only after this the question of the nature of consciousness may be posed and resolved. It turns out that the brain does not produce consciousness but is rather an instrument of consciousness. Important processes (first of all super-intuition), that are starting and finishing in consciousness, are performed nevertheless in the unconscious state.quantum coherency is achieved in these processes because they deal with the quantum reality, i.e. with the whole quantum world. The obstacle of decoherence does not appear in this case because the quantum world as a whole has no environment that could cause decoherence.

Therefore, starting from functions rather than material carriers of functions turns out the only efficient approach. One of the astonishing conclusions is that some functions have no concrete material carriers or, alternatively, have the whole world as their carrier. This leads in fact to the unification of the sphere of material with the spiritual sphere.

•

The idea that this approach may be fruitful appeared during the preparation of the review at the famous Ginzburg's seminar in Moscow. The aim of the review was the novel applications of quantum mechanics called quantum information. However, this issue is closely connected with the foundations of quantum mechanics. In the process of work on this topic it unexpectedly occurred to me that the main features of consciousness including its mystical abilities are explained if a simple logical construction is added to conventional quantum mechanics. Especially exciting was that this additional assumption actually simplified the logical structure of quantum mechanics.

This was astonishing and led to further investigations that revealed the deep interconnection between the concept of quantum mechanics and the phenomena characteristic for life. It turned out that mysterious character of life explains those features of quantum mechanics that are counter-intuitive and vice versa. The most deep theory of inanimate matter expressed in the form of quantum mechanics supplies just those notions and abilities that are necessary for understanding of the (otherwise mysterious) phenomena of consciousness and life.

The central role in this internal connection is the so-called "quantum reality". This counter-intuitive concept was investigated in various ways beginning from the famous Einstein–Podolsky–Rosen paradox and ending by Everett's interpretation.

The Everett's picture of actually quantum world as the set of many coexisting parallel worlds (alternative classical realities) expresses the concept of quantum reality in the most transparent way. If one thinks about consciousness, keeping in mind that actual reality is not a single classical world but many equally real (although subjectively seeming to be alternative, excluding each other) classical worlds (as alive and dead Scrödinger cat), he/she understands what is consciousness including its mystical features (super-intuition, or direct vision of truth, and even how one may "manage reality").

This conclusion appeared unexpectedly, but actually it has been prepared by the long history of insights of genius physicists into the internal sense of quantum mechanics. It seems that now we are also close to the better understanding of what is quantum mechanics. It is exciting that this new level of understanding is directly connected with the phenomena of life and consciousness.

Michael B. Mensky
December 2009
Moscow

Acknowledgments

The author is grateful to his colleagues for many discussions providing the high level of understanding the most complicated ideas of quantum theory of measurement, foundations of quantum mechanics and especially the Everett's interpretation of quantum mechanics. My supreme gratitude is to Dieter Zeh and Vitaly Ginzburg.

Many long discussions with Professor Zeh in his hospitable house in the neighborhood of Heidelberg gave me the very deep idea both of decoherence and of Everett's "Many-Worlds" interpretation of quantum mechanics. Professor Ginzburg requested for me to give a series of talks on quantum measurements and quantum information at his famous seminar in Moscow. After this he invited me to publish the reviews in the Russian review journal "Physics–Uspekhi". Moreover, he initiated the discussion of this "eternal" topic in the journal. This gave me strong motivation for more active research in this direction that finally implied Extended Everett's Concept treating consciousness in the context of quantum mechanics. The present book is an account of the results of this research at the present day.

Michael B. Mensky
November 2009
Moscow

Contents

Miracles produced by consciousness (psychic experience)

25

Parallel worlds and consciousness

37

List of Figures

Chapter 1

Introduction: From quantum mechanics to mystery of consciousness

"For the invisible reality, of which we have small pieces of evidence in both quantum physics and the psychology of the unconscious, a symbolic psychophysical unitary language must ultimately be adequate, and this is the far goal which I actually aspire. I am quite confident that the final objective is the same, independent of whether one starts from the psyche (ideas) or from physis (matter). Therefore, I consider the old distinction between materialism and idealism as obsolete."

Wolfgang Pauli

(From the letter by Pauli to Rosenfeld of April 1, 1952. Letter 1391 in [Meyenn (1996)], p. 593. Translated by Harald Atmanspacher and Hans Primas in [Atmanspacher and Primas (2006)].)

In this chapter we shall briefly enlist the main ideas of the book and their origins. In the following chapters this list of ideas will be filled by the concrete contents, and the logic which makes these ideas convincing will be traced.

Some of the following chapters will be written with the usage of special terminology and mathematical apparatus of quantum physics. They will be marked as intended for readers-physicists and may be skipped without detriment for understanding the main line of consideration.

1.1   Questions to be answered

There are questions that cannot be answered (or at least cannot be convincingly answered) in the context of modern science:

- What is consciousness?
- What is unconscious and why is it so important?
- Is it possible to find out truth intuitively, if no information for this is available (super-intuition)?
- Is foresight of the future possible?
- Is it possible to manage reality, i.e. influence the events by the power of the consciousness?
- Can consciousness "create miracles"? Are miracles actually incompatible with natural sciences?
- Can the phenomenon of life be reduced to physical and chemical laws or is there something else in this phenomenon?
- Why living beings are so efficient in surviving?
- How health is supported in an organism and why the most dangerous diseases sometimes disappear without any medicine?
- Is it possible to overcome the global crisis of our technical civilization?
- What is the nature of the great scientific insights?
- Can natural sciences (including quantum physics) be purely objective and ignore subjective elements (consciousness of observers).
- How the work of a scientist should be organized at the moment when a novel view on the problem is necessary for its solution (i.e. how a scientist can initiate the scientific insight)?

These questions will be considered in the present book together with the natural scheme of consideration providing answers to all of them. This system may be called theory of consciousness and unconscious. It will shed light not only on the phenomenon of consciousness (mind), but also the phenomenon of life.

All these questions are from the area of spiritual life of humans or, more generally, concern the mystery of life. It turns out that such questions can be completely or partially answered if the specific understanding of the concept of reality, unavoidable in quantum mechanics, *quantum reality*, is taken into account.

We shall consider quantum reality and theory of consciousness, taking, as a starting point, the so-called *Many-Worlds interpretation of quantum mechanics* suggested in 1957 by Hugh Everett.

Although quantum reality and Everett's interpretation may be presented in all details only for professional physicists, most of the important ideas in this area may be presented in a simpler form available for non-professional people. We shall provide the simplest possible approach to the problem, demonstrating the essence of this approach by examples, metaphors and graphical illustrations. However, in each case we shall give also the strict quantum-mechanical consideration (although in simplest possible formulation).

1.2 Two spheres of knowledge

There are two spheres of knowledge (spheres of cognition) which are quite distinct.

- One of these spheres is the natural sciences that deal with the objectively existing material world and its laws. The scientific laws are in essence simple and concern simple (or rather elementary) objects such as elementary particles. Technically complicated calculations arise (when applying the fundamental laws to real situations) as secondary effects caused by a large number of elementary objects and arbitrary initial and boundary conditions. The scientific laws are expressed in terms of the sophisticated mathematical apparatus and are analytic, i.e. are aimed at the reduction of complicated systems to their elementary components.
- Another sphere concerns the sphere of subjective experience of a person, his/her consciousness. This sphere includes knowledge of the rich internal world of a human. The contents of this sphere are expressed in the form of the images and ideas together with their verbal expressions (sometimes long texts) rather than short formulas. Conclusions in this sphere are typically based on synthesis rather than analysis.

These two spheres of knowledge seem to have nothing in common, since their methods, subjects of the investigation and the very nature of their contents are different. Nevertheless, there is a very important connection between them. It exists because each of these spheres turns out to be in a sense incomplete (for example logically unclosed) if the other sphere is excluded from the consideration.

- The deep analysis of the sphere of spiritual life of human reveals such aspects of this sphere that are directly connected with the work of consciousness and arise feeling of something mysterious, not yet understood or even not understandable. These aspects are conveniently called mystical. If they are considered separately from the natural sciences, then the consideration of mystics and of the whole spiritual sphere seems to be naive, out-of-date, taken from the past and having no roots in the present. However, attempts to understand or explain human consciousness and especially its mystical features from the viewpoint of the natural sciences give no convincing results.

- The sphere of natural sciences looks (and actually is) modern, deeply rooted, well-substantiated and reliable. However, the deep analysis of its logical structure clearly demonstrates that the very core of this knowledge which lies in the area of quantum physics contains conceptual problems, or paradoxes. These problems cannot be solved until the second of the two knowledge spheres (spiritual sphere) is explicitly accounted. At least the consciousness of an observer has to be included into the consideration for the description of measurement be complete in quantum mechanics.

We shall consider in this book the approach to unification of these two spheres of knowledge on the basis of the conceptual structure of quantum mechanics. The main issue will be the interpretation of the phenomenon of consciousness in terms characteristic for quantum mechanics. This is not derivation of consciousness from quantum physics. It is rather constructing theory of consciousness starting from the ideas invented for solving the internal conceptual problems of quantum mechanics.

The conceptual problems of quantum mechanics become evident in description of measurements (observations) of quantum systems. The origin of these problems is in the specific concept of reality accepted in quantum mechanics. may be formulated Therefore, quantum theory of measurement and the concept of quantum reality will serve as the starting point for theory of consciousness.

The logical chain leading from quantum mechanics to theory of consciousness begins in the necessity to include the observer's consciousness as a necessary element in theory of quantum measurements. It is important that the expansion of quantum mechanics due to this necessity leads finally not only to solving the internal problems of the quantum mechanics itself, but also to understanding what is consciousness, thus giving contribution to the spiritual sphere of knowledge.

Due to the specific character of quantum reality, purely objective science turns out to be impossible. The subjective component of our knowledge must be necessarily accounted. The nature of our world may be completely explained only on the basis of the unification objective (natural-scientific) and subjective (mental, or spiritual) spheres of knowledge. The unification of these so different areas should conserve the richness of each of them as well as their relative independence of each other.

1.3   Super-intuition: Where do right solutions come from?

Anyone knows about the efficiency of intuition. It provides right solutions of the most complicated problems. It is often supposed that intuition is only the ability to think very rapidly, deriving the conclusions with the help of the usual rational arguments by very quickly, almost instantaneously. However, intuitive solutions are available even in the situation when there is no rational background for such solutions.

We shall use the special term, *super-intuition*, to underline this specific situation when the right solution is found although there was no way to logically derive it from the information available in the usual way.

Super-intuition is in a sense obtaining information that seemingly cannot be obtained. This mystical ability "to make what cannot be made" is nevertheless actually observed. To explain why this is possible will be one of our tasks.

What is the basis for super-intuitive solutions? Where the information for such a solution come from if no information is available by the conventional means? We shall argue that all this is possible due to *direct vision of truth*, the special ability of our consciousness. Quantum mechanics explain why this is possible.

1.3.1   *Super-intuition in life and in science*

You surely know the situation when you have to accept an important decision, but, just because of its vital importance, cannot choose one of several options. The indeterminacy may continue for a long time, often many days, causing painful feeling of helplessness and despair. It is impossible to stop continuous fruitless thinking on the problem that again and again goes in the same circle of reasoning but gives no result. How to stop this end-less thoughts, how to choose one of a number of solutions avoiding fatal mistakes?

The answer is amazingly simple. You should briefly survey your rea-soning once more and completely stop thinking on this problem. In order to take mind off the subject, it is helpful to make something pleasant, may be simply go to cinema or theater. The decision will come unexpectedly, accompanied by the delightful feeling that it is the only right one. Fu-ture experience confirms that this decision is in fact the best of all possible options.

Here are two bright examples of such situations.

A diver who is going to make a record of the depth of submergence without apparatus experiences great danger at the moment when he/she reaches the maximum depth and turns backward. He/she has to choose the moment of returning in such a way that to achieve as deeply as possible but have enough time to achieve the surface. Returning a little bit later may mean death. How to make right decision in this critical situation? The experienced sportsmen (sportswomen) tell that before this critical moment they sink into a sort of trance and make the choice of the returning moment in the unconscious state.

The other example occurred with Russian cosmonaut Grechko. He ex-perienced the off-nominal situation on his return on the Earth in one of his cosmic trips. The min engine was down and he had to turn on the small subsidiary engine with restricted resource. Then he had to turn it off in the manual mode so that his cosmic apparatus began slowly descend in the regime of free fall. wrong choice of the moment of turning the engine off could lead either to heavy landing or to staying the apparatus on the orbit without chance to land. Grechko had no way to calculate the right time, but he chose it intuitively and avoided both dangers. The choice was made in great emotional tension, and it is most probable that the cosmonaut was in the state of trance.

How and why this happens? Why right solutions of the most important problems are found instantaneously and without any grounds for these so-lutions? The short answer is that the decision is chosen in these case by intuition. However, the well known word 'intuition' denotes in this case a strange ability of our consciousness, the ability of direct vision of truth. The intuitive solution of the problem happened to be in this case valid just because it has been not a simple guess but the direct vision of truth.

The same phenomenon takes place also in case of "scientific insights" when an unexpected solution of a scientific problem (or a principally new direction of thinking on the problem leading to the solution) is found not by rational reasoning but as a simple guess having no logical ground. Of course, the guess of this type comes only after the scientist was systematically working on the problem by usual rational methods and thus clearly formulated the problem in the scientific terms.

1.3.2 *Parallel alternatives (parallel worlds): what does this mean*

Very briefly, consciousness and super-consciousness (usage of super-intuition) may be explained by parallel worlds predicted by quantum mechanics. This is reflected in the title of the present book.

Someone asked me: "Life in parallel worlds... Who lives there - in these parallel worlds?"

Many people write nowadays about "parallel worlds", meaning various things behind this term, but mostly some modification of oriental beliefs. One psychic talks about four "worlds", describing in detail how they look, what are their constructions, who lives there and what are these worlds for. He said even how each of the worlds is called. I asked him how can he know about all this, especially about names of the worlds. He answered that one of his pupils (each year he is teaching psychic practice for a group of young people) is regularly traveling along these worlds and tells him about them.

Of course, I mean not this. Logic of quantum mechanics leads to such conclusions that it is difficult to believe in them but it is impossible to ignore them. Among these conclusions, the most important is that the quantum world, with its "quantum reality", may be adequately presented as the set of many classical worlds, *parallel worlds*. These classical worlds are in fact the various "projections" of a single objectively existing quantum world. These classical worlds differ from each other by some details, but they are pictures of the same quantum world. These parallel classical worlds coexist, and we are parallelly living in all of them (a clone of each of us in each of these classical worlds).

Thus formulated, the concept of many coexisting classical worlds is counter-intuitive. And it is counter-intuitive, but only from the point of view of classical intuition. In quantum mechanics it cannot be otherwise. The reason is that for any given classical state of a quantum system[1] its

---

[1] More precisely, almost classical. A quantum system cannot be in a purely classical state, but some of its states are close to classical, almost classical.

future state is presented as a number of coexisting (superposed) classical states. At the next step each of these classical states converts into the set of a number of coexisting (superposed) classical states and so on. The result is the enormous number of parallely existing (superposed) classical states.

This argument is applicable to the whole quantum world which is also (infinite) quantum system. Therefore, typical state of the quantum world is the set of enormous number of parallel classical worlds.

To agree this strange picture (which is in fact confirmed by many experiments) with the everyday experience, physicists suggested that of all possible alternative classical worlds arising with time, a single one is randomly chosen in each moment, so that always only a single world exists. However, this suggestion, convenient as it may be, is in fact incompatible with the strict logics of quantum mechanics. As a result, the *well-known paradoxes of quantum mechanics.*

It is only in 1957 (i.e. three decades after the quantum-mechanical formalism had been created), a young American physicist Hugh Everett III turned out to be bold enough to consider such an interpretation of quantum mechanics according to which no choice of a single worlds is made, so that all parallel worlds do actually coexist.

The interpretation of quantum mechanics that accepts objective coexistence of many distinct classical worlds has been called Everett's, or Many-Worlds interpretation. Not all physicists believe in this interpretation, but the number of its adepts is increasing rapidly.

The Everett's worlds which have to coexist due to the essential nature of quantum mechanics (due to "quantum concept of reality") are the *"parallel worlds"* considered in this book. We see a single world around us, but this is only illusion of our consciousness. Actually all possible variants (alternative states) of this world coexist as Everett's worlds. Our consciousness percepts all of them, but separately from each other: subjective feeling of the perception of one of the alternative worlds excludes any evidence of the others.[2]

1.3.3 *Consciousness and quantum mechanics*

The essence of the Extended Everett's Concept (EEC), or Quantum Concept of Consciousness (QCC) suggested by the author and considered in this book is that turning the consciousness off (as in sleep, trance or med-

---

[2]One may say that we live in *Alterverse*, the set of parallely existing alternative classical worlds, or alternative classical realities. This term is an analogue of the term "Multiverse" used in quantum cosmology for the set of many quantum Universes.

itation) exclude separation of the Everett's worlds from each other. Then all of them together are available for what is left instead of consciousness and what can be called *super-cognition* because it supplies the information that is not available in the usual (conscious) state. Turning consciousness to the process of super-cognition and backward to consciousness can be called *super-consciousness.*

The super-consciousness provides the access to all variants of evolution of the world we see around us (alternative scenarios of the world) and can find out what of them is advantageous. This gives the unique information unavailable by the usual perception organs and explains the phenomenon of super-intuition, or "direct vision of truth". In more common situations (when consciousness is turned on but some processes in the organism are regulated in unconscious regime) this explains the mechanism of surviving (support of health), i.e. the very phenomenon of life.

One more natural suggestion is that super-consciousness may not only obtain information from the whole set of Everett's worlds but influence the probability of each of them to be subjectively felt in future. This gives a mechanism for influence on *"subjective reality"* and may explain *"probabilistic miracles"*, i.e. seeming violation of scientific laws. Actually no laws are violated in this case, but the probabilistic (stochastic) nature of quantum-mechanical laws is exploited.

All this may look complicated at the first glance, but is in fact very simple and natural in the context of quantum mechanics in its Everett's "Many-Worlds" interpretation (which in turn is the only logically closed interpretation). The whole EEC (or QCC) includes, in comparison with the original Everett's interpretation, only two additional assumptions. One of them explains the phenomenon of surviving ("miracle of life") and super-intuition (direct vision of truth). The other explains "probabilistic miracles", i.e. the ability to arbitrarily choose the subjective reality by consciousness and super-consciousness.

This book is devoted to the mentioned abilities of our consciousness and many other relevant phenomena. Some of them are known as mystic events, some are similar to miracles (the special type of miracles, connected with the consciousness and unconscious).

The phenomena of this type are investigated by various spiritual traditions including various religions, oriental philosophies, esoteric doctrines, parapsychology etc. However, we shall consider them from the scientific point of view.

At the first glance, the phenomena of this type contradict to modern natural science and are impossible from the scientific viewpoint. This however is not valid if such mysterious branch of science as quantum mechanics is taken into consideration. Moreover, it turns out that quantum mechanics is logically incomplete and needs theory of consciousness to be included in it for becoming logically closed. The quantum concept of reality is such that the resulting theory of consciousness (and unconscious as an essential element) predicts quite unusual abilities of consciousness, among them direct vision of truth and "probabilistic miracles".

The thought that quantum mechanics and consciousness are closely connected has been stated by many authors, beginning from Wolfgang Pauli in collaboration with Carl Jung and up to Roger Penrose. During the long history of quantum mechanics important new aspects of the connection between quantum mechanics and consciousness were analyzed and the efficient mathematical instruments for this were developed. It is now almost evident that the so-called Many-Worlds interpretation of quantum mechanics (Everett's interpretation) should play the key role in the final theory connecting consciousness with quantum mechanics.

The ideas of Pauli concerning this issue were not widely known until the end of 20th century, because Pauli never published them, discussing this topic only in letters to his friends. Now his short thoughts on this subject arise great interest and are often cited and discussed (see for example [Atmanspacher and Primas (2006); Enz (2009)]).

The Extended Everett's Concept (EEC) proposed by the present author in 2000 connects the issues of consciousness and quantum mechanics by a very short chain of reasoning. This makes the resulting theory quite plausible. In this book we shall present and develop this scope of ideas, trying to do this in the simplest possible way.

Remark 1.1. The following two remarks have to be added about the usage of the term "consciousness".

- This term, as it is used in literature, is not quite unambiguous. By conscious one may mean various psychic phenomena. Everywhere in this book we understand this term in the sense originating from the quantum-mechanical term "consciousness of an observer". This sense of the word may be defined as the most deep and at the same time most primitive aspect of the phenomenon, the "root of consciousness". This is what differs the state "I aware that I perceive something" from the state when nothing is perceived and the person is not aware of anything. Contrary to this, the word "consciousness" is often understood as denoting intellectual processes developing on the background of consciousness (for examples calculations or rational thinking).[3]

- The state of unconscious plays the key role in all phenomena discussed in this book (including the phenomenon of the superintuition mentioned above). In fact, the most important for these phenomena is the interrelation between the states of consciousness and unconsciousness. Therefore in many cases, talking about the whole scope of the discussed phenomena and saying for example "the role of consciousness" we shall mean the role of the states of conscious and unconscious and transitions between these states.

1.4 Principle of life is not derived from but is added to science

This book is about the phenomena of consciousness and life and their explanation on the basis of quantum mechanics. The task to explain these phenomena is very old, and the task to explain them with the help of quantum mechanics is very popular nowadays. Yet the approach taken in this book radically differs from what other authors suggested.

Usually the scientists, in their attempts to explain consciousness and life, tried to derive the phenomena of life and consciousness from the laws of motion of matter. In other words, they tried to reduce these phenomena to the laws found by such sciences as chemistry and physics. This direction of research may be called *reductionism*. Despite of many interesting achievements on this way, this approach never gave positive results in the main goal of reductionism: in reducing the laws of living matter to the laws found in the investigation of the inanimated matter.

New hopes to obtain such an explanation was connected with the new ideas of quantum mechanics, such as quantum information and quantum computers. Usually the hypothesis is considered that some structures in brain work as a quantum computer. However, no significant results were achieved in this direction too. Quantum version of reductionism does not work too, although the hopes connected with it do yet exist.

---

[3]We shall almost never make use of the word "consciousness" in this sense. The exclusion is the task mentioned in Sect. 9.1 to change consciousness from egotistic to altruistic.

The approach suggested by the author in 2000 and exposed in the present book is different. According to this approach, the phenomena of life and consciousness cannot be mechanistically reduced to the action of the laws of science as they are found in the course of exploring the inanimated matter. The explanation of these phenomena on the basis of quantum mechanics requires addition of a special independent element to the set of quantum concepts and laws.

Such a new element of theory should directly connect quantum concepts with the concepts characteristic of life. The simplest way to find this element is to consider the phenomenon of consciousness and compare it with the description of observation (measurement) in quantum mechanics. Then it may be formulated as identification of consciousness with the "separation of the alternatives" — a concept relating to the "Many-Worlds" interpretation of quantum mechanics. It is interesting that the addition of this element simplifies the conceptual structure of quantum mechanics instead of doing it more complicated.

If we consider not only the phenomenon of consciousness but more general phenomenon of life, this additional element may be called *"life principle"*. It very naturally follows from the analysis of theory of consciousness, but in fact it acts for all forms of life, even simplest forms having no consciousness. The life principle formulates evolution of living system in such a way that it is determined by the goals as well as by causes. The main goal of the living system is survival so that their evolution provides their survival. However, for more sophisticated forms of life, the goals may include other criteria of the quality of life.

The phenomena of life and consciousness therefore cannot be reduced other to quantum mechanics or to any other theory of inanimated matter. Of course, the laws or these sciences act in the processes performing in the bodies of living organisms, but life and consciousness are not the direct consequence of these processes. Life is not the function of a body, and consciousness is not a function of the brain. Rather body is a realization of life, and brain is an instrument of consciousness.

Life and consciousness are something additional to the natural sciences, even additional to quantum mechanics. Yet the main features of life and consciousness (including the most deep, mystical features of them) are naturally connected with the specific feature of quantum mechanics called "quantum reality". This is why life and consciousness can be understood on the basis of quantum mechanics. In order to guess what are the main points of theory of life and consciousness, one can start from quantum mechanics and analyze the most deep, counter-intuitive features of quantum mechanics, those which make this branch of science strange and not quite transparent.

The idea of the additional assumption that should be accepted to go over from quantum mechanics to theory of consciousness is hinted by the conceptual structure of quantum mechanics itself. This is the approach applied by the present author to find the explanation of consciousness and life. The secret of this approach that gave very interesting results is very simple: one has to analyze the conceptual structure of quantum mechanics, first of all its conceptual problems (paradoxes) forgetting all the dogmas, explicitly or implicitly existing in science. Then, on the way to the most simple formulation of the structure of quantum mechanics, the additional assumption is suggested, that simultaneously 1) simplifies the conceptual structure of quantum mechanics and 2) explains the phenomenon of consciousness.

The simplicity of the resulting logical construction and important consequences following from it give the impression that the correct way is found. The results may then be analyzed from various points of view including philosophical ones.

1.5 Graphic presentation of the relation between the two spheres

Carl Jung compared the relation between the sphere of psychic and the material world with the two cones having a single common point coinciding with the vertex of each of them (Fig. 1.1).

Fig. 1.1 Two spheres of knowledge have common point, special for each of them.

We shall show later that the common point (or rather the common area of the two spheres of knowledge) is nothing else than the concept of consciousness as well as the circle of concepts and phenomena related to consciousness. It is important that all the concepts and phenomena in this common area are not, up to now, well understood in the framework of natural sciences. We shall argue that the interpretation of them as belonging to both spheres provides their explanation.

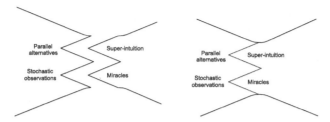

Fig. 1.2   Left picture: Quantum mechanics (left) has defects, or paradoxes; spiritual knowledge (right) includes mystical features. Right picture: if both spheres of knowledge are joined, the paradoxes of quantum mechanics explain the mystical features of spiritual knowledge.

This may be symbolically presented in Fig 1.2. This figure takes into account two paradoxical features of quantum mechanics: *parallel alternatives* (paradoxical because alternatives cannot parallely coexist in classical physics) and *stochastic nature of the results of observations* (paradoxical because in classical physics an ideal observation with a given initial state is unambiguously determined, its results cannot be stochastic). The figure expresses symbolically that these two paradoxical features in quantum mechanics provide the explanation of the corresponding mystical features experienced in the psychic practice: *super-intuition* (direct vision of truth) and *probabilistic miracles*. This will be discussed in detail later in the book.

Thus, the connection of the two spheres of knowledge is performed in the area corresponding to the special (paradoxical) concepts in the natural sciences and special (mystical) phenomena in the spiritual knowledge. The special area of the natural sciences is connected with quantum paradoxes (quantum reality). The special area of the spiritual knowledge is mystics and miracles related to psychic (consciousness).

It is natural that the configuration presented symbolically in Fig. 1.2 provides more clear understanding of the concepts of both spheres of knowledge than it is possible without their confronting. The theory unifying both material (natural-scientific) and spiritual (mental, psychic) knowledge will better explain what is "consciousness" (mind, psychic).[4]

---

[4]The approach based upon quantum mechanics will show that mystical features of consciousness appear when permanent or temporary transition to the regime of unconscious occur. Therefore, the term "phenomenon of consciousness" includes in fact the interrelation of conscious and unconscious states of "mind". Talking of "nature of consciousness", we often mean the nature of the phenomena reflected somehow in our consciousness but actually depending on both conscious and unconscious states.

1.6 Toward theory of consciousness

It may seem strange but consciousness, the phenomenon which is well known to each one, is not understood by modern science. If consciousness is a product of brain, then theory of consciousness should be elaborated in the framework of neurophysiology. And actually physiologists sometimes claim that they understand what is consciousness. However, despite of radical improvement of the technology applied in the research, physiology cannot explain the nature of consciousness as such (of course, good progress is made in the investigation of the intellectual processes being realized on the background of consciousness as such).

The lack of success in the explanation of consciousness shows that the nature of consciousness cannot be understood in the limits of chemical, physical or information processes realized in the brain. This is indirectly confirmed by strange phenomena observed in consciousness and phenomenologically denoted as mystical. It is almost evident that mystical features of consciousness hardly could be explained as a result of physical and chemical processes in brain.

However, while consciousness cannot be understood in the context of chemistry, classical physics and physiology, it turns out that it (or at least its main features) can be understood in the context of the quantum mechanics. More precisely, the essence of consciousness can be interpreted as a special type of perception of *quantum reality* by living beings.

1.6.1 *Mystical features of consciousness are compatible with quantum mechanics*

Mysticism and mystical features of consciousness were treated long before emergence of the modern science, in various types of pre-scientific knowledge. However, nowadays the scientific explanation of any phenomenon is expected. If something is observed but not explained by natural sciences, it is usually considered as not confirmed. Therefore, the question of relation between mystical features of consciousness and natural sciences is actual.

Mysticism includes miracles, and this seems to exclude its scientific explanation. Indeed, a *miracle* simply by definition is something that is cannot exist in reality. In a more precise formulation, a miracle is something that, according to the laws of natural sciences, cannot occur. Is not it evident that this exclude mystical phenomena from the scope of those existing in reality? Strangely enough, but this "evident" conclusion is incorrect. The phenomena looking as mystical can be observed, and this does not contradict science.

The explanation of this paradoxical statement is in *probabilistic nature of quantum-mechanical laws*.

If reality were described by classical physics, mystical phenomena could not exist in reality. However, after great scientific revolution of the first quarter of 20th century we know that *reality is correctly described only by quantum physics*, and only approximately it can be presented by classical equations. Precise laws of nature are quantum, and one of the cardinal difference of quantum laws is their probabilistic, or stochastic, nature.

This feature of quantum laws is revealed when quantum system undergoes measurement. Even if the state of the system before a measurement is precisely known, the result (readout) of the measurement cannot be unambiguously predicted. It is possible to enumerate the alternative measurement results and predict probability of each of them. Such a probabilistic law can be verified only by a long (ideally infinite) series of measurements. Those alternative measurement results which are more probable should happen more often, less probable measurement results should occur rarely.

But this means that *a single measurement can neither confirm nor refute any probabilistic law*. Let one of the possible measurement results has very low probability, say 10^{-9}. Almost all people, including professional physicists, will consider observing this result "practically impossible". According to this, observing this result of measurement in reality "practically contradicts" to the given law, so that this observation would be a miracle.

However, considering the situation in the mathematically strict way, we can only predict that in an extremely long series of measurements (many millions of events) the given result will be observed on the average in one event from each million of events. However, it cannot be predicted in what concrete measurements this result will be observed. It may be observed even in the very first measurement of the series, and this would not contradict to the probabilistic law. Moreover, this measurement result, although having very low probability, may well occur if the measurement is performed only once. This happening, strange as it may look, would not contradict the probabilistic law.

The final conclusion is in fact astonishing: *a single event may look as a miracle, without any contradiction with a probabilistic quantum-mechanical law*. Quantum mechanics allows strange events that can be called *probabilistic miracles*.

Thus, the phenomena that look as miracles (i.e. mystical phenomena) are compatible with the modern natural sciences, because quantum mechanics is the heart of these sciences and probabilistic miracles are allowed by it. This principal possibility is realized in Quantum Concept of Consciousness (QCC), the theory of consciousness, following from quantum mechanics. We shall discuss it briefly in the next sections and in more detail in the subsequent chapters.

1.6.2 *Quantum mechanics is incomplete without consciousness*

We are going to explain consciousness, including its mysterious features, on the basis of modern science, because they badly need a sort of scientific explanation. It turns out however that science also needs inclusion consciousness in its structure. The reason is that *quantum mechanics is logically incomplete without inclusion the concept of consciousness*. Quantum physicists do not often aware of this because the mathematical structure of quantum mechanics, including the probabilistic laws, is quite correct. This provides correctness of all calculations and solution of all practically arising problems. However, when deeply analyzed, quantum mechanics is met with *conceptual problems (paradoxes)* that cannot be solved without inclusion of subjective element, for example the concept of consciousness.

The conceptual problems of quantum mechanics are revealed in description of measurements (observations) of quantum systems (shortly, in *quantum measurements*). They may be also illustrated in a transparent form as paradoxes.

1.6.2.1 *Paradox of Schrödinger's cat*

To illustrate paradoxical character of quantum mechanics (existing conceptual problems in it), one of the creators of this branch of science, Erwin Schrödinger, suggested the following thought experiment. In fact this paradox illustrates the difference of the concept of reality in quantum mechanics from the reality as it is meant in classical physics and in our usual intuition.

Take a black box and put into it a cat together with an unstable (gradually decaying) atom and an automatic device destroying an ampule with poison if the atom is decayed. Then at the beginning of the experiment the atom is not decayed and the cat is alive. If at some moment the atom is decayed, then the cat is dead. These two cases are clear and do not differ from what can exist according to classical physics. However, the atom, as a microscopic object, obeys quantum mechanics, and this implies unusual conclusions.

According to quantum mechanics, any state of any quantum system is a vector. This means that, just as in case of the usual vectors, the states of the quantum system may be summed up.[5] The result of summing two or several state vectors are called in quantum mechanics *superposition*.

The state of the atom at the initial moment is "non-decayed", but with time it becomes the superposition (non-decayed atom + decayed atom), with the first term of this sum gradually decreasing and the second increasing.[6]

Let us recall now that the state of the cat is directly connected with the state of the atom. We have to conclude then that the state of the system atom+cat is (non-decayed atom and alive cat + decayed atom and dead cat), see Fig. 1.3.

Fig. 1.3   Schrödinger's cat in quantum superposition. "Quantum reality" suggests coexisting the parallel worlds (alternative classical realities) such that the cat is alive in one of the worlds and it is dead in another world.

Now what shall we see if opening the black box at this moment? Can we see the cat in the state describing as a superposition of the alive cat and the dead cat? Evidently not. We shall alternatively see either the alive cat (and yet non-decayed atom) or the dead cat (and already decayed atom).

This is the paradox. Describing the state in the closed box according to quantum mechanics we have to present this state as the superposition. But for the open box the description, in accordance with our experience, should be one of the component of this superposition.

We see that in this reasoning, leading to a paradox, *essential role is played by our consciousness*. Until the box is open, the information about

---

[5]They may be also multiplied by (complex) numbers, but this is not important for us at the moment.

[6]This means that the first (correspondingly second) term is multiplied by the increasing (correspondingly decreasing) coefficient.

the state of the system did not yet enter into our consciousness, after opening the box we are conscious of this state.

The main conclusion from the paradox of Schrödinger's cat (that is in fact a simplified model of a more general situation of a quantum measurement) is necessity (in the context of quantum mechanics) of superpositions even for macroscopic systems, such as a cat (or measuring device). This requires *serious revision of the concept of reality, which finally lead to the theory of consciousness.*

1.6.2.2 *Quantum reality*

Let us say a few words about a *quantum measurements*, situation that generalizes the situation of Schrödinger's cat.

The main conclusion from the consideration of quantum measurements is following. In quantum mechanics superpositions of states may exist (when states are summed up as usual vectors). This is proved by enormous number of experiments with microscopic objects. However, consideration of quantum measurements shows that superpositions of states of macroscopic systems should also exist.

A superposition may include (as its components) *macroscopically distinct states*, such as alive and dead cat or the state of the measuring device with the pointer directing to the left and another state with the pointer directing to the right. Such superpositions cannot be identified with anything emerging in practice of observers (a cat either alive or dead, but not both, the pointer directs to the right or to the left, but not both). This is one of the characteristic features of what is called *quantum reality*. The seemingly contradiction of this from the observations needs a special explanation. Such an explanation is given in the Many-World interpretation of quantum mechanics suggested in 1957 by Hugh Everett. The following steps lead from the Everett's interpretation of quantum mechanics to the "quantum theory of consciousness".

1.6.2.3 *Many-Worlds interpretation of quantum mechanics includes consciousness*

Thus, logically we have to conclude that not only microscopic objects, but also macroscopic objects are also quantum and therefore may be in the states of superpositions. Moreover, the components of the superposition may be macroscopically distinct: alive and dead cat in the Schrödinger's cat paradox, the pointer of the measuring device directing to the right and to the left as a result of a quantum measurement.

This contradicts to our everyday experience (but more precisely—to the experience of our consciousness). This is the reason why this straightforward conceptual conclusion of the basics of quantum mechanics was not accepted for decades after completion of its mathematical formalism. However, in 1957 the bold enough physicist made this simple step: Hugh Everett III proposed his famous *Many-Worlds interpretation of quantum mechanics*.

According to this interpretation, any states of our (quantum) world may coexist as the components of superposition. These coexisting states may be macroscopically distinct. We are used to think, on the ground of the experience of our consciousness, that coexisting of macroscopically distinct states of the world is impossible. However, this proves possible because quantum mechanics requires this, and quantum mechanics is very well verified.

To make the situation more transparent or better compatible with our common-sense concept of reality, the physicists suggested another terminology: not the different states of the quantum world coexist, but different classical worlds coexist as the components of the superposition. The only objectively existing quantum world is a superposition of different classical worlds, often called *Everett's worlds*. Thus, in the situation of Schrödinger's cat paradox the objectively existing quantum world is a superposition of the two classical Everett's worlds. In one of these Everett's world the cat is alive, in the other Everett's world the cat is dead.

We shall accept another terminology that may be less transparent but much more convenient for the analysis. We shall say that (in the above example) the objectively existing quantum world is *objectively* in the state of the superposition of two states (the alive and dead cats coexist, usually alternative to each other), but our consciousness perceives the components of this superposition separately from each other, or *the consciousness separates the alternatives*. This means that an observer may see the alive cat, but then he/she does not see the dead cat, and vice versa. Both alternatives objectively coexist, but separated in consciousness (subjectively).

1.6.3 *Theory of consciousness from quantum mechanics*

Everett's interpretation of quantum mechanics allowed to overcome the conceptual problems (paradoxical nature) of this science. However, much more important is that this interpretation allows to do the next step. This interpretation allows *to understand what is consciousness* and explain its strange, unbelievable, but nevertheless practically observed mystical features.

As has already been said, it is necessary (for compatibility with our everyday experience) to assume that the alternatives (Everett's worlds) *are separated in consciousness*. The present author proposed to do one more step and *identify consciousness with separation of alternatives*. The resulting theory was called *Extended Everett's Concept* (EEC). It may be also called *Quantum Concept of Consciousness* (QCC). This concept explains the nature of consciousness (not explained otherwise) in terms of quantum mechanics.

If the first step (identifying consciousness with the separation of alternatives) is done, an important next step may naturally be done: we can conclude something about *unconscious* (which is known to be very important for human psychic). Indeed, if conscious is the separation of the alternatives, then turning consciousness off is the disappearance of the separation. Therefore, in the state of unconscious the alternatives (all Everett's worlds) are somehow accessible all together, without any separation between them. Remark that they are not perceived in the usual sense of the word (because the usual images are impossible in the unconscious regime), but yet somehow reflected.

It is important that the information about all these (parallel worlds) is available so that they can be compared with each other and the most favorable of all these alternatives can be found. The information about what alternative is the best (favorable) is the basis for *super-intuition*, or direct vision of truth. This wonderful phenomenon that seems to be observed in practice finds thus its explanation in the special features of quantum mechanics.

Another assumption that seems natural in the context of QCC, or EEC, is that consciousness can modify *"subjective probabilities"* of various alternatives. Then those alternatives (Everett's worlds) that are favorable may be made subjectively more probable even if their objective probabilities are very small. The resulting phenomenon may look as a miracle, as the managing reality. However, this is subjective reality rather than objective one. The phenomenon may be called *probabilistic miracle* and turns out to be quite compatible with the probabilistic laws of quantum mechanics.

This line of consideration, leading finally to the main points of the unified theory of matter and spirit, will be followed in detail in the rest of this book. We shall try to present the material parallely on the two different levels: first, in a simple form, available for any reader, second, for professional physicists, in a more sophisticated and more professional form, with more details and more areas of research included. The chapters or sections including the complicated material will be indicated as being intended for physicists.

PART 1
Miracles produced by consciousness (psychic experience)

Science became a sort of religion in the 20th century. At the same time, the expansion of mysticism in various forms is an evident tendency in the modern society. This is strange because the central point of mysticism is admittance of miracles, which seem to be impossible, according to natural sciences. Most part of scientists deny mysticism as contradicting scientific viewpoint. However, the evidences confirming miracles (at least those created by consciousness) became now more numerous and better documented. We shall survey here some spiritual schools which accept mystics. Besides, we mention typical examples of the strange phenomena, similar to miracles, that are produced by consciousness.

In the subsequent parts of the book, it will be shown that the events of this type, strange as they might seem, belong to the special type of events, called probabilistic miracles, and are not in real contradiction with science. Moreover, such special and in fact mysterious science as quantum mechanics cannot be considered logically complete without consciousness being included as its counterpart. This results in explanation of many very deed sides of the phenomenon of consciousness including super-intuition and probabilistic miracles

Chapter 2

Miracles and mysticism in spiritual experience of mankind

Consciousness as the very important, perhaps the most important ability of a human being, has been explored a long time earlier before science (in the modern sense of the world) appeared. The exploration of consciousness in those ancient times was concentrated first of all on its unusual, mystical features. This was the central point of all spiritual doctrines, most of which survived, in some form or another, up to our time.

The aim of the present book is investigation of the phenomenon of consciousness in the context of quantum mechanics. We shall show that the sphere of spiritual knowledge is not independent of and not contradict to the modern science. In the present chapter we shall briefly survey the main spiritual traditions that treated, by their specific means, the mystical features of the phenomenon of consciousness. One of the goals of this survey is to demonstrate that these old doctrine hardly may be considered troglodytic as some people think. Vice versa, they express, by their specific ways, those aspects of the phenomenon of consciousness that cannot be adequately studied by the methods of modern natural sciences.

This chapter will serve then as a starting point demonstrating importance of the coinvestigation of spiritual and scientific concepts.

2.1 Historical background

Ancient knowledge, often including elements of mystic, survived during centuries and is popular even now, in the epoch of science. Moreover, the popularity of some of these non-scientific interpretations of our world increases, often accepting new forms. One of the reasons of this increasing interest to the ancient forms of knowledge is in the increasing number of facts confirming that our consciousness possesses unusual abilities interpreted as mystical. We shall discuss later these facts from the scientific position, but here we shall very briefly survey various doctrines and directions of thinking accepting mysticism.

2.1.1   *Religion*

Religion is a great part of the human culture originated in remote ages and giving rise ourdays to a number of world confessions. Many people acknowledge belief in God, some of them believe also in Biblical miracles. Many scientists also accept God, but their belief is less naive, more abstract. Einstein said that he believed in the "God of Spinoza", or "God who reveals Himself in the harmony of all that exists."

If a scientist considers himself believer, he in a sense separates his science from his belief so that they do not interfere with each other. The question of the scientific view on the mystic aspect of religins is thus overcome without solving it. Those scientists who consider this question seriously were rather rare, although now it becomes more popular with each year.

The great physicists Wolfgang Pauli was working, in collaboration with the great psychologist Carl Gustav Jung, on the connection between quantum mechanics and the misterious phenomena in consciousness. He has published no paper on this topic and expressed his thoughts about this only in letters to the colleuges. Now a number of books about the Pauli-Jung collaboration and their views were published. Much more will be published on the topic in the next years. We are now enough prepared for reading and thinking about this.

2.1.2   *Oriental philosophies*

Oriental philosophies are the most amazing achievements intermediate between religion and science. They amy be considered to realize the experimental scientific approach to the investigation of the subjective sphere of human and its relation to the objective world. The experimental method in this sphere must be based on observing work of consciousness (mind). The thousands of years of such experiments gave enormous amount of practical knowledge in the form of practical recommendations and abstract doctrines resembling both religion and philosophy.

Practical achievements of the oriental people are well known. They look as ability to create miracles, mostly concerning functions of their bodies. Typical is the widely known experiences with stopping breath after the proper preparation of psychic. Analogous experiments on "dancing on fire" are practiced even more widely, not only at East, but also in South Europe. I myself have seen "dancing on fire" in mountain Bulgarian village, while my friend, professor of philosophy, was even a participant of such an action in Siberia, and told me about the severe psychic preparation of the participants.

It is astonishing that important aspects of the phenomenon of consciousness as it is treated in oriental philosophies are quite similar to those following from quantum mechanical analysis. We shall see this later.

2.1.3 *Esoterica*

Esoteric knowledge is that which is available only to a narrow circle of "enlightened", or "initiated", people. Esoteric items may be known as esoterica. (In contrast, exoteric knowledge is knowledge that is well-known or public, or perceived as informally canonic in society).

There are various branches of esoterica. Esotericism is not a single tradition but a large number of movements, having no single historical thread underlying them all.

Several historically attested religions emphasize secret or hidden knowledge, which may be called esoteric. Some saw Christianity, with its ritual of baptism, as a mystery religion. The terms "Gnosticism" and "Gnosis" refer to a family of religious movements which claimed to possess secret knowledge (gnosis). Another important movement from the ancient world was Hermeticism or Hermetism. Both of these are often seen as precursors to esoteric movements in the scholarly sense of the word.

Western esoteric movements in the scholarly sense have roots in Antiquity and the Middle Ages. A major phase in the development of Western esotericism begins in the Renaissance. Pursuits of Antiquity that entered into the mix of esoteric speculation were astrology and alchemy. A second major source of esoteric speculation is the kabbalah, which was lifted out of its Jewish context and adapted to a Christian framework. Outside the Italian Renaissance, yet another major current of esotericism was initiated by Paracelsus, who combined e.g. alchemical and astrological themes into a complex body of doctrines.

In the early 17th century, esotericism is represented by currents such as Christian Theosophy and Rosicrucianism. A century later, esoteric ideas entered various strands of Freemasonry. Later in the 18th century, as well as in the early 19th century, the diffuse movement known as Mesmerism became a major expression of esotericism. In the 19th century, esotericism is also represented e.g. by certain aspects of the philosophy, literature and science associated with Romanticism, by spiritualism, and by a notable French wave of occultism.

The major exponent of esotericism in the latter part of the 19th century is the Theosophy of H. P. Blavatskaia. In the 20th century, Theosophy was reformulated and became the source for a whole range of post-theosophical movements. A particularly successful post-theosophical movement is Anthroposophy, which includes esoteric versions of education, agriculture, and medicine.

Yet another notable esoteric strain stems from the teachings of G. I. Gurdjieff and P. D. Ouspensky.

Finally, it can be noted that Carl Gustav Jung, can be seen as an exponent of esotericism: his writings concern esoteric subject such as alchemy, and rephrased the concept of correspondences in a modern, psychologizing terminology in his theory of *synchronicity*.

2.2   Psychic and parapsychology

2.2.1   *Edgar Cayce*

2.2.1.1   *General data*

Edgar Cayce (1877 1945) was an American who was psychic. He demonstrated an ability, while in a self-induced trance, to answer what is the disease of a given person and how he/she may be made health. Cacey is known also as a prophet who predicted World War Two, but we shall concetrate on his abilities to heal people that were unique and well documented.

Cayce's healing practice was utterly successful. Most of his *readings* were given for the people living far from him and known to him only by the names. The vast majority of his readings allowed to overcome the problems of the patients. It is very important that most of the Cayce's readings and their consequences were accurately documented by the special commission which included professional physicians. This makes the methods and results of his healing practice authentic.

Edgar Cayce Centers are now found not only in the United States and Canada, but also in 25 other countries. The Association for Research and Enlightenment (ARE), headquartered in Virginia Beach, Virginia, is the major organization promoting interest in Cayce.

2.2.1.2 Details of the practice

The Cayce's practice was organized in the following way. On the receiving a letter with the request of healing Cayce lied down on a couch and sank into a trance and began what he called reading. Knowing only the name of the patient, he began to mutter: "This body is now in the town..." He named the town and the position in which the body was at the moment, then described the physical state and the health problems of the person. Then he spoke about the measures that could help in this case. The measures might be of various kind: the known or less known medicines, the diet, the physiologic or medical procedures, and so on.

The measures, taken on the Cayce's advices almost always helped. If not, once more reading was arranged and additional measures adviced. Sometimes those who were besides the patient could not find the necessary potion. then, in a new readings, Cayce advised where it may be taken. It happened that Cayce pointed out the concrete pharmacy (sometimes in abother town) where the necessary drug is available. In one case the host of the pharmacy named by Cayce could find the required drug. Then, in the special reading, Cayce pointed out the concrete shelf where, behind the other bottles, a bottle with the necessary but flong-forgotten potion stood.

The special features seen in the readings of this unique psichic are following: in the state of trance he could see the people and subjects located far from him and find out the inner state and capabilities of each one. It is especially astonishing that he knew precisely how each measure will act, as if he followed into the future the picture of what happens if one or another measure is taken.

The activity of Cayce was confirmed by the special commission that was observing him for many years. This provides an amazing evidence of quite unusual abilities of the human consciousness. the details of the phenomenon of Cayce will be very important later when we shall discuss the phenomenon of consciousness (mind) in the context of quantum mechanics.

2.2.2 Health by the autosuggestion

It is known that a disease may sometimes be overcome (at least partly) with the help of suggestion or autosuggestion, therefore, by the . power of consciousness. In the course of suggestion or autosuggestion the claims of the type of "You are quite healthy", "Your heart is working perfectly", or "I am quite health", "My hart is working perfectly" are repeated many times and finally change the actual state of the organism. The more sophisticated variant of such a method is auto-training when the suggestion is performed in the state of the deep relaxation.

It seems strange that the health may be restored simply by the power of fantasy. However, this proves to be possible. Moreover, there is an opinion that the very deep reasons of most of diseases are violations in the work of consciousness.

Many people, even if they believe in this strength of the consciousness, estimate it as less valuable than the action of the usual medical procedures. Often, if a person feels better as a result of suggestion or autosuggestion, they say scornfully: "This is only the autosuggestion", meaning that this is much worse than medical treatment. But is it right?

This viewpoint originates in fact from the conviction that human cannot be healthy without proper medical treatment. But this is evidently wrong at least for young people. An young one is healthy without any care at all. The human organism therefore possesses all that is necessary for being healthy, including the means for recovery in case of departure from the norm. There are many evidences that the primary reason of the age-related diseases is not the deficit of the facilities necessary for recovery but rather inefficient regulation in their application.

It is well known that the usual diseases are very rare during war. It is known that the ill mother instantaneously becomes healthy if her baby happens to be in danger. Therefore, a soldier and a woman have everything necessary for being healthy, but these means are properly applied when the soldier is in the conditions of war and the woman needs all her strength to help her child. To be healthy, one needs therefore not special chemical substances but proper functioning of the systems of his organism.

Examples of a soldier at war and mother defending her child demonstrate the consciousness may provide the proper functioning of the organism. The suggestion or autosuggestion may help this, and often this is more efficient or less dangerous than application of traditional (mostly chemical) medical treatment.

We shall argue later that the very deep and in fact necessary condition for health is the action of consciousness (or rather super-consciousness, the ability existing in the state of unconscious). From this point of view it is not at all strange that suggestion and autosuggestion may be efficient for the restoration of health.

2.2.3 Telepathy, clairvoyance etc.

Various phenomena of parapsychology were known (or may be believed to exist) long time ago, but became popular in the middle of 20th century. It seems natural that such phenomena as telepathy, clairvoyance, foresights etc. were believed in Middle Ages, but why the believe in them became so common in the scientific 20th age looks strange. Some people consider that the reason of this is in hard state of psychic of most people in the modern world that makes psychologically desirable easy (magic) solutions of hard problems. This well may be one of the reasons. However, it is difficult to ignore one more and much more rational reason: the appearance of many well documented facts of strange mystic phenomena.

The most known are the evidences of foresights, i.e. predictions. Some of them (such as the foresights of Nostradamus) are formulated in the rather unclear form and cannot in fact serve for confirming the very ability of provisions. However, the other (such as predictions made by Casey of Bulgarian psychic Vanga) are documented and followed up to the realization of the predicted events.

There are other types of mystic phenomena: telepathy and clairvoyance. Many people meet the evidences of such phenomena in their experience or the experience of the people around them. The most usual example is mystical connection between close relatives, for example mother and son. for example, a mother may feel the very moment of the unexpected death of her son. The evident explanation of such facts by simple coincidences is always possible but seems insufficient because of many events of this type that occur in dramatic situations.

The evidences of mystic phenomena should not be considered doubtful if they are very often or well documented. Nevertheless, most of people and at least most of scientists simply ignore these facts. Why?

The 20th century changed the status of science making it in fact a sort of religion. Application of the scientific methods for verification of data are considered to be the only reliable and universal criterion of these data being truth. The mystic phenomena seem to be in evident contradiction with science, so they are usually rejected by those who consider themselves educated. However, gradually serious doubts appeared in the universal character of the scientific criterion of truth (at least in the common sense of the word "scientific").

The mystic phenomena are always connected with consciousness, and the nature of consciousness is not clearly understood in the modern science.

Therefore the science in its modern state cannot be the ultimate judge in the area of these phenomena.

We shall see in the following that both the nature of science and the abilities of consciousness looking as mystic phenomena are not only compatible with science but are even necessary in order to overcome the internal contradictions existing in science, namely in quantum mechanics, the most powerful but simultaneously mysterious branch of science.

2.3   Miracles in science: Scientific insights

We live in the epoch when science achieved great success and became an universal origin of knowledge about the world. Before appearance of science the means of investigation of world were of quite other nature. The main difference is that the ancient knowledge arose in the form of statements (for example of priests of the predominated religion) not supported by any systematic procedure of proof. This is why modern people often consider ancient knowledge to be naive and unreliable in comparison with the scientific knowledge. However, is this right?

In science the laws of nature are elaborated by the systematic investigation of the observed phenomena. They are proved to be valid with the help of series of experiments. However, the most important steps in this process look as unexpected insights not at all following from any logical line of reasoning. Of course, the hypotheses found in this way have to be conformed later by regular scientific methods (experiments and logical reasoning). However, no nontrivial laws of nature (such as special or general relativity or quantum mechanics) could be formulated without illogical insights made by geniuses (such as Albert Einstein) at the key moments of the development of science. Then the difference between ancient wisemen and modern geniuses in science is only in that the latter are supported by the army of professional but not genius scientists making technical work.

This argument makes clear why the ancient knowledge is not abandoned after appearance of the modern scientific knowledge but occupies its own place in the human culture. Moreover, the ancient knowledge proved its genuineness by the very fact of its surviving during centuries.

The modern science is commonly believed to be quite opposite to the ancient non-scientific knowledge. It turns out that the basic knowledge may be found only by a sort of foresight or direct vision of truth. The ability of such direct vision of truth was always available for human beings. We shall discuss later, on the basis of quantum mechanics, why direct vision of truth is available for human beings. The difference of the knowledge provided by the modern science in comparison with the ancient forms of knowledge is in high and systematic development of terminology and systematic application of the procedures of proof.

Yet the comparison, very interesting on the fundamental level, is not simple and not always correct in details and concrete applications of knowledge now and in remote ages. The subject of the modern science, aiming at the technological tasks, differs from the subject of the ancient knowledge aiming at the general features of existence. Natural sciences nowadays formulate fundamental laws of motion of simple forms of matter in rather simple situations. More complicated situations arising in applications are regulated not by fundamental laws as such but should be calculated on the basis of fundamental laws by fantastically developed mathematical methods. Nothing of this existed in old time.

Anyway it is more important not to compare the modern science with the ancient knowledge that accepted mysticism and miracles but to understand what is the status of the mysticism from the viewpoint of the natural sciences. The traditional opinion is of course that they contradict each other. However, we shall see that deep analysis discover no contradiction between science and some features of consciousness that appear as mystical phenomena. Moreover, the most successful but also mysterious branch of science, quantum mechanics, naturally leads to theory of consciousness predicting such features.

PART 2

Parallel worlds and consciousness

"Quantum reality" taking place in our world is coexisting (parallely existing) alternative classical realities (parallel worlds): all worlds that could in principle exist, do actually exist. All objectively existing worlds are perceived by our consciousness, but separate from each other. Perceiving one of these worlds is accompanied by the illusion that there is no other.

Yet actually existing all the other parallel worlds implies mystical features of our consciousness: *super-intuition* (access to the information from "other parallel worlds") and *probabilistic miracles* (increasing the probabilities to subjectively perceive those parallel worlds which are advantageous). The "probabilistic miracles" do not contradict natural sciences because probabilistic behavior is a fundamental feature of quantum physics.

Outlined thus way of reasonings, which leads from quantum mechanics to the explanation of the phenomenon of consciousness, is called Extended Everett's Concept (EEC) when it is presented for the physicists with the support on the appropriate mathematical formalism. The simplified account of the same approach for the wider audience is called Quantum Concept of Consciousness (QCC). The more general consideration, including the general phenomenon of life and not directly referring to the phenomenon of consciousness, is called Quantum Concept of Life (QCL).

Those readers who are not physicists may read only Chapter 4 skipping Chapters 3, 5, 6 intended mostly for professional physicists.

Chapter 3

Quantum reality as parallel classical worlds (for physicists)

"In their beds at night, children ask for details about a fairy tale. How big was the pumpkin? What color were Puss-in-Boots' boots? In the same way, our reason questions our positive understanding. Now then, all that physics! Does it really disclose nothing but rules and recipes?"

Bernard d'Espagnat In Search of Reality [d'Espagnat (1983)]

"We are here only in the very beginning of a new development of physics, which will certainly lead to still further generalizing revisions of the ideals underlying the particular description of nature which we today call the classical one."

W. Pauli, Dialectica 2 (1948), p. 311

Most of the present chapter is written for those who dealt with quantum physicists. The reader having no knowledge about this specific branch of science may skip it without detriment for understanding further chapters.

We shall demonstrate in this chapter why quantum mechanics needs explicit introduction of the notion of consciousness and more generally of subjective elements in it. This branch of physics cannot be made logically closed without this radical step first made by Everett in 1957 in his famous "Many-Worlds" interpretation of quantum mechanics. The essence of this interpretation may be formulated as the assumption that many macroscopically distinct classical realities coexist despite that they are alternative (excluding each other) from the usual point of view. Equivalent formulation is in coexisting parallel worlds (Everett's worlds).

This counter-intuitive concept of parallel (Everett's) worlds supplies an adequate formulation for "quantum reality", the notion that first appeared in the famous Einstein–Podolsky–Rosen paradox and was further clarified in the phenomenon of quantum non-locality, Bell's theorem and Aspect's experiments. We shall show in the subsequent chapters that quantum reality, or parallel worlds naturally lead to the deep understanding of the phenomenon of consciousness including its mystical features.

3.1 Introduction

3.1.1 *Consciousness and quantum mechanics: From Pauli and Jung to contemporary authors*

Attempts to understand the phenomenon of consciousness from the point of view of quantum mechanics may be followed back up to the collaboration of Wolfgang Pauli and Carl Jung at the first quarter of 20-th century. The results of this collaboration have never been published by the authors completely and only in our days became familiar to the wide audience (see for example [Enz (2009)] and the references therein).

Not knowing the thoughts of Pauli and Jung, many other people attempted to find the bridge connecting quantum mechanics and consciousness (see for example [Squires (1994)], [Lockwood (1996)], [Whitaker (2000)], [Stapp (2001)], [Penrose (2004)], [Zeh (2000)]). It is evident now that the work in this direction should be based upon the interpretation of quantum mechanics suggested by Everett [Everett (1957)] and elaborated further by other authors.

We shall expose in the present book the approach to this problem proposed by the present author [Mensky (2000a)] and elaborated in a series of papers and the book in the further years. This line of consideration has been called Extended Everett's Concept (EEC). It seems the most convincing and leading to the most interesting conclusions.

The present chapter is in fact introductory in respect to this program. It will present the so-called "measurement problem" of quantum mechanics. This term denotes the complex of conceptual problems of quantum mechanics, appearing in the description of measurements of quantum systems.

Just these problems make inevitable the appearance of the concept of "consciousness of an observer" and opens further road to the explanation of the human consciousness. However, we shall discuss these issues later, restricting ourself now by the internal problems of quantum mechanics connected with quantum measurements.

3.2   An observer's consciousness and quantum paradoxes

3.2.1   *Special features of quantum measurements*

One feature of quantum measurements is that a quantum system cannot be measured (i.e., any information about it gained) without perturbing its state, and the more information is extracted in the measurement, the stronger the perturbation. This, of course, is well known and is quantitatively treated typically using the uncertainty relation. [1]

It is also known that even with the exact knowledge of the state of a system it is usually impossible to predict the measurement result with certainty.[2] Generally, it is only possible to calculate the probability distribution over various measurement data.

This is quite sufficient for practical purposes. Predictions based on the probabilistic calculations allow to achieve all practical aims, and quantum system measurements present no problems in this sense.[3]

This 'trouble-free' approach is theoretically formulated in terms of a quantum ensemble of similar systems in the same state. Knowing the probability of one measurement result or another, we know what fraction of the systems that make up the ensemble yield a given result in the measurement. In the general case, we are not allowed to know more; the quantum-mechanical predictions of measurement data (or of observations) are no more than probabilistic in nature.

3.2.2   *Paradoxicality of quantum mechanics*

By adopting this ideology, one can successfully work in quantum mechanics and never encounter the notorious 'measurement problem'. Does this mean

---

[1]Let us remark, by the way, that there are some measurement-related subtleties in the seemingly quite familiar uncertainty relations that are ordinarily neglected, see Ch. 3 in Ref. [Mensky (2000b)] about this.

[2]Definite predictions available only in exceptional cases, when the system prior to the measurement is in one of the eigenstates of the observable being measured.

[3]We shall see however (see Section 4.4), that this is very important outside the pure physics, because it makes some mysterious features of consciousness compatible with physics.

that there is no problem at all? No, quantum mechanics has the problems not yet solved, but they are of special nature. These are conceptual problems. This is why they are sometimes formulated in the form of paradoxes, the famous *quantum-mechanical paradoxes*. This is, for instance, the paradox of Schrödinger's cat (see Section 1.6.2.1). One more is the paradox of Wigner's friend.

Physicists of a practical mind are not interested in paradoxes as long as the problems they have to solve are well-posed. It is unreasonable though to completely forget about the paradoxes, on which such outstanding physicists as the authors of the above-mentioned paradoxes Schrödinger and Wigner, as well as Einstein, Bohr, Heisenberg, Pauli, Wheeler, DeWitt, and many others considered it necessary to spend their time and energy.

The paradoxicality of what takes place in a quantum measurement, as it is demonstrated by the Schrödinger's cat paradox, is further emphasized in the paradox of Wigner's friend.

Wigner [Wigner (1961)] considered a situation in which not he himself but his friend performs measurements of some quantum system, and then, after the measurement is completed, lets him know the measurement result. The result eventually reduces to the fact that the system is in one of two states: $|\psi_1\rangle$ or $|\psi_2\rangle$. These states are called 'pointer states' because they correspond to the alternative measurement results. The experimenter learns about the state of the system from whether he sees a light flash in the corresponding measuring device (the light flash being in the case an equivalent of the position of a pointer). As in the paradox with the Schrödinger's cat, in this case, too, prior to the measurement the system is in a state that is a superposition of the states $|\psi_1\rangle$ and $|\psi_2\rangle$ (say, $c_1|\psi_1\rangle + c_2|\psi_2\rangle$). But the crucial question is how should we describe the state in which the system resides after the measurement.

It turns out that the description of the final state of the system, just as in the Schrödinger's cat paradox, depends on the observer's consciousness. If the experimenter has not yet looked at the device, he describes the state as the superposition of the states $|\psi_1\rangle$ and $|\psi_2\rangle$. If he has, then either as $|\psi_1\rangle$ or as $|\psi_2\rangle$ (depending on precisely what he has seen). The description of the system state depends on whether the experimenter has become aware of the system state.

3.2.3 *Wigner friend paradox*

We have already seen this (in Section 1.6.2.1) in the paradox of Schrödinger's cat. But Wigner introduces a new element because his experimenter-friend conveys to him, Wigner, the information about the measurement. For as long as Wigner does not possess this information, he describes the system state as a superposition of $|\psi_1\rangle$ and $|\psi_2\rangle$. On receiving the information, he describes it differently: either as $|\psi_1\rangle$ or as $|\psi_2\rangle$ (depending on the contents of the information transmitted). Therefore, the description of the state of Wigner's system depends on whether his consciousness has perceived the information about the measurement result that his experimenter-friend has transferred to him.

The paradoxicality of the situation is underscored by the following reasoning. Wigner says, "However, if after having completed the whole experiment I ask my friend, "What did you feel about the flash before I asked you?" he will answer, "I told you already, I did [did not] see a flash," as the case may be. In other words, the question whether he did or did not see the flash was already decided in his mind, before I asked him."

3.2.3.1 *Entanglement*

To realize clearly what is odd about this, we translate it into the language of formulas.

Let the prior-to-measurement state of the system under measurement be

$$|\psi\rangle = c_1|\psi_1\rangle + c_2|\psi_2\rangle,$$

and the state of the device be Φ_0. Then, the state of the compound system (comprising the system to be measured and the device) prior to the measurement is given by the state vector (wave function)

$$|\psi\rangle|\Phi_0\rangle = (c_1|\psi_1\rangle + c_2|\psi_2\rangle)|\Phi_0\rangle.$$

Let Φ_1 denote the state of the measuring device in which a flash occurs and Φ_2 denote its state in which no flash occurs. Then, the measurement result perceived by the observer is described by either the vector $|\psi_1\rangle|\Phi_1\rangle$ (if he sees the flash) or $|\psi_2\rangle|\Phi_2\rangle$ (if he does not see it). The former signifies that the device has transited from the Φ_0 state to the Φ_1 state, while the system under measurement has found itself in the $|\psi_1\rangle$ state. The latter is interpreted in a similar manner.

The states of a compound system (comprising two subsystems) like $|\psi_i\rangle|\Phi_i\rangle$ are said to be factored. This means that each of the subsystems is in a definite (pure) state, i.e., is characterized by a state vector (wave function). We can assume, however, that the measuring device has already been actuated but the observer has not yet looked at the device. Then, the state of the complete system (including the system under measurement and the device) is obtained from the initial state

$$(c_1|\psi_1\rangle + c_2|\psi_2\rangle)|\Phi_0\rangle = c_1|\psi_1\rangle|\Phi_0\rangle + c_2|\psi_2\rangle|\Phi_0\rangle$$

by the action of the linear evolution operator or the solution of the linear Schrödinger equation. This necessarily, simply due to the linearity of this operation, yields

$$c_1|\psi_1\rangle|\Phi_1\rangle + c_2|\psi_2\rangle|\Phi_2\rangle.$$

This state is not factorisable. We cannot in this case point out a definite state vector (wave function) for each subsystem. Instead, the subsystems are said to be *entangled*. One may say that one of the subsystem is in the state $|\Phi_i\rangle$ *if* the other is in the state $|\psi_i\rangle|$. In other words, the states of the two subsystems are correlated. This is so-called *quantum correlation* (or *entanglement*) qualitatively differing from the correlations that exist in classical physics (see Section 3.4).

3.2.3.2   *Final conclusions*

Let us formulate the conclusion of our consideration.

(1) Insofar as the observer has not become aware of the measurement result, he is guided exclusively by quantum-mechanical laws and should therefore describe the state of the complete system by the vector

$$|\Psi\rangle = c_1|\psi_1\rangle|\Phi_1\rangle + c_2|\psi_2\rangle|\Phi_2\rangle$$

(entangled state). Once he has realized the measurement result, he describes the state by one of the vectors

$$|\Psi_1\rangle = |\psi_1\rangle|\Phi_1\rangle \quad \text{or} \quad |\Psi_2\rangle = |\psi_2\rangle|\Phi_2\rangle;$$

depending on precisely which result he observes.

(2) Wigner describes the state by the $|\Psi\rangle$ vector insofar as his friend has not let him know the measurement result, but after the announcement, by one of the vectors $|\Psi_1\rangle$, $|\Psi_2\rangle$.

(3) Once Wigner's friend (the experimenter) answers the question "What did you feel about the flash before I asked you?," Wigner should draw the following conclusion: even prior to receiving the message but knowing that the measurement has taken place and his friend knows the

measurement result, he, Wigner, has to describe the state by one of the vectors $|\Psi_1\rangle$, $|\Psi_2\rangle$ (not knowing by which of the two, though). In this case, Wigner's description of the state is determined by his knowledge of the fact that his experimenter-friend has looked at the device, i.e., the consciousness of his friend has perceived the information about the measurement result.

Yet another subtlety emerges when we consider the situation where there is no live observer (Wigner's friend) of the device. In this case, simply by the linearity of quantum-mechanical equations, Wigner (like any other physicist in this situation) must describe the after-measurement state by the vector

$$|\Psi\rangle = c_1|\Psi_1\rangle + c_2|\Psi_2\rangle.$$

If the 'measuring instrument' is microscopic, for instance an atom, additional experiments may allow verifying (from the presence of interference effects) that the correct state description is indeed provided by the vector $|\Psi\rangle$ rather than $|\Psi_1\rangle$ or $|\Psi_2\rangle$. In the case of a macroscopic device, simply because of technological reasons, there is no way of carrying out such a verification, but the vector $|\Psi\rangle = c_1|\Psi_1\rangle + c_2|\Psi_2\rangle$ may be derived theoretically, relying exclusively on the linearity of quantum-mechanical equations (for instance, the Schrödinger equation).

All this led Wigner to conclude [Wigner (1961)] that a living observer plays a special part in quantum mechanics, somehow breaking the linear nature of evolution. When the information about the result of a measurement (observation) enters the observer's consciousness, the state description becomes such that it cannot result from the evolution described by a linear operator.

Wigner's paper was written a long time ago, back in 1961, and at first sight its arguments seem to be naive. But in reality, they reveal deep and truly specific features of quantum measurements, which are fully comprehended from a purely formal, mathematical aspect but do not get along well with our intuition. The conclusion from the above that is most significant for the subsequent discussion is that the observer's consciousness should be explicitly taken into consideration in the analysis of a quantum measurement. This can also be substantiated in other ways.

3.3   Reduction and decoherence in a measurement

3.3.1   *Reduction*

A quantum measurement may be formally represented with the aid of a procedure called *state reduction* or *wave function collapse*. Reduction is closely connected with the phenomenon of *decoherence*. The presentation of quantum measurements in terms of reduction and decoherence is in good agreement with our intuition, and this is the reason why this presentation is commonly accepted. It is actually very important for understanding the relation between quantum and classical descriptions of what happens in measurements. Let us consider this circle of concepts.[4]

In the simple example given in the previous section, the initial state $|\psi\rangle = c_1|\psi_1\rangle + c_2|\psi_2\rangle$ experiences reduction during measurement; as a result, it passes into the state $|\psi_1\rangle$ with the probability $|c_1|^2$ and into the state $|\psi_2\rangle$ with the probability $|c_2|^2$. [5] The state reduction (in combination with similar procedures that follow from the reduction and describe more complicated measurements) provides the correct phenomenological description of a quantum measurement.

3.3.2   *Entanglement*

The question involuntarily arises of what 'really' takes place in this case and how so strange a transformation of the state as its reduction occurs. A partial answer to this question is provided by the phenomenon of *entanglement* and, as its consequence, *decoherence*. We shall briefly characterize these phenomena by taking advantage of the example in Section 3.2.

As we have seen in the foregoing, when we consider the measuring device as some quantum system and apply a conventional quantum-mechanical description to its interaction with the system under measurement, the result of the interaction between these two systems is that their initial state

$$|\Psi_0\rangle = (c_1|\psi_1\rangle + c_2|\psi_2\rangle)|\Phi_0\rangle$$

---

[4]Later on we shall see that this description is logically incompatible with the linearity of quantum mechanics and shall find the proper place for it in the conceptual structure of quantum mechanics.

[5]More generally, according to von Neumann's reduction postulate [von Neumann (1932)], every (perfect) measurement is characterized by a complete system of orthogonal projectors $\{P_i\}$, and with the ith measurement result, the initial state of the system $|\psi\rangle$ passes into $P_i|\psi\rangle$.

passes into the state

$$|\Psi\rangle = c_1|\psi_1\rangle|\Phi_1\rangle + c_2|\psi_2\rangle|\Phi_2\rangle.$$

The state of the form $|\Psi_0\rangle$ is said to be '*factorized*', because it is represented by the product of subsystem state vectors. The state of each of the subsystems in this case is characterized by a certain state vector. The after-measurement state $|\Psi\rangle$ belongs to the class of *entangled states* of two subsystems (in this case, the system being measured and the device). The two subsystems in an entangled state are said to be *quantum-correlated*.

3.3.3 *Decoherence*

Let us go over from entanglement to the phenomenon of *decoherence* [Zeh (1970); Zurek (1981, 1982); Joos and Zeh (1985); Giulini et al. (1996)] (see also the book [Mensky (2000b)] of the present author).[6]

An entangled state cannot be represented as a product of two state vectors pertaining to the subsystems (cannot be factorized). This signifies that although the compound system comprising both subsystems is in a pure state (i.e., its state is represented by a state vector, in this case, $|\Psi\rangle$), the subsystems considered separately are not in pure states (i.e., cannot be represented by state vectors). Instead, each of the subsystems can be individually characterized by a *density matrix*.

For the system under measurement, the density matrix is found as follows:

$$\rho = \mathrm{Tr}_\Phi\left(|\Psi\rangle\langle\Psi|\right) = |c_1|^2|\psi_1\rangle\langle\psi_1| + |c_2|^2|\psi_2\rangle\langle\psi_2|,$$

In this calculation, to the density matrix $|\Psi\rangle\langle\Psi|$ of the combined system, we have applied the operation of partial trace over the states of the system Φ (i.e., the device). Under this operation, there emerge scalar products $\langle\Phi_i|\Phi_j\rangle$ of the basis states of this system, and when the states $|\Phi_1\rangle$ and $|\Phi_2\rangle$ are orthogonal and normalized, the expression written on the right-hand side follows.

The density matrix, unlike the state vector, describes not a pure state but what is called a mixed state. The mixed state can be interpreted as the probability distribution over some set of pure states. In this instance, the density matrix signifies that the subsystem resides in the pure state $|\psi_1\rangle$ with the probability $|c_1|^2$ and in the pure state $|\psi_2\rangle$ with the probability

---

[6]Decoherence attracted great attention and has been analysed from various viewpoints in the work of a range of physicists. Besides the above-dited authors, we can mention Murray Gell-Mann, Jim Hartle and Stephen Hawking.

$|c_2|^2$. It is easily seen that this corresponds to the ordinary probabilistic description of a quantum measurement, i.e., to the reduction postulate: the measurement may yield the former result with the probability $|c_1|^2$ (and then the system being measured is in the state $|\psi_1\rangle$) and the latter result with the probability $|c_2|^2$ (with the system in the state $|\psi_2\rangle$).

The transition of the pure state $|\psi\rangle$ to the mixed state ρ is termed *decoherence*, because it is accompanied by the loss of information about the relative phase[7] of the complex coefficients c_1 and c_2. In the example considered, the decohering of the subsystem resulted from the interaction of this subsystem with another subsystem. The interaction leads to entanglement of both subsystems, and this means that each of them decohered.

Therefore, when we want to describe, after the measurement, only the system being measured and not include the measuring device into the description, we must use the density matrix rather than the state vector and mixed states rather than pure ones. It is significant that the density matrix is derived by conventional quantum-mechanical techniques and contains the probability distribution over different measurement results.

If we are concerned only with probabilistic predictions (and this would be quite sufficient for all practical purposes) and have no need of any deeper analysis, the density matrix and the decoherence effect it represents may be thought of as providing the complete picture of a quantum measurement. There is nothing paradoxical about this picture and no problems like the 'measurement problem' arise at this level of analysis.

But we now revert to the deeper level of analysis. We take advantage of the approach proposed by John Bell, which has come to be very popular because presented the conceptual problems of quantum mechanics in a transparent and experimentally falsifiable form.

3.4   Quantum correlations and quantum reality

Existence of *quantum correlations*, or *entanglement*, of two (or more) subsystems of a quantum system is a specific feature of quantum mechanics. Nothing of this type exists in classical physics. Correlations exist also in classical physics, but they are very simple for understanding. Quantum

---

[7]The pure state $|\psi\rangle$ can also be represented by the density matrix $\rho_0 = |\psi\rangle\langle\psi|$. If ρ_0 is expressed in terms of the vectors $|\psi_1\rangle$ and $|\psi_2\rangle$, it turns out to differ from ρ by the presence of non diagonal terms proportional to $|\psi_1\rangle\langle\psi_2|$ and $|\psi_2\rangle\langle\psi_1|$. That is why decoherence is also defined as the disappearance of non diagonal terms in the density matrix.

correlations form such a novel phenomenon that it is not easy to completely understand it. In fact, comprehension of this phenomenon has been achieved, after a very long time period, through a number of crucial points such as Einstein–Podolsky–Rosen (EPR) effect, or paradox, Bell's theorem and Everett's interpretation of quantum mechanics. The result of this long process of gradual comprehension is concept of *quantum reality*.

3.4.1 EPR effect and Bell's inequalities

It is extremely significant that the features of quantum measurements are impossible to explain (to resolve the paradoxes) in any logically simple way. For instance, one might endeavor to attribute the probabilistic nature of predictions of measurement results to the absence of complete information about the initial state. In other words, one might assume that in the measurement of a quantum system, everything proceeds just as it does in the measurement of a classical system, with the difference that we do not know the initial state of the system exactly and cannot therefore predict the measurement results precisely. However, this assumption proves to be incorrect. The fallacy in this assumption is clearly demonstrated by *Bell's theorem* [Bell (1987, 1964)] and experiments like Aspect's experiment [Aspect et al. (1981); Aspect (1982)], which rule out 'local realism'. This signifies the following.

Bell's inequalities emerge in the analysis of experiments of the *Einstein–Podolsky–Rosen* (EPR) type, proposed in the famous paper Ref. [Einstein, Podolsky and Rosen (1935)]. The most clear form of an EPR-type experiment has been proposed in 1951 by David Bohm. In this (thought) experiment a zero-spin particle decays into two particles with spins 1/2, and the spin projection on some axis is measured for each of the produced particles.

These measurement data are correlated in a specific manner. This is clear from the mere fact that the sum of the spin projections of all particles participating in the reaction is conserved. This sum is equal to zero prior to the decay and should therefore remain zero after the decay. The correlation is evident when measurements for two particles are made of the projections on the same axis. Then, when the projection is equal to $+1/2$ for the first particle, the projection for the second particle turns out to be $-1/2$, and vice versa.

When the axes along which the spin projections are measured do not coincide, the correlation is more complicated but is inevitably present (with the sole exception of orthogonal axes, when the correlation vanishes completely).

John Bell considered the implications that would emerge if the spin projections had specific values prior to their measurements or at least the particles prior to the measurement could be characterized by some probability distribution of their spin projections on given axes. The existence of a probability distribution of this type, even prior to the measurement, is characteristic of classical physics and has come to be known as *'local realism'*.[8] Bell showed that the EPR measurement data should, under the assumption of local realism, necessarily satisfy certain inequalities, which are referred to as *Bell's inequalities*.

Therefore, having carried out the measurements and checked if Bell's inequalities are satisfied, one can verify the validity of local realism. When Bell's inequalities are not satisfied, the assumption of local realism is to be rejected.

Calculating the probabilities of different measurement data according to quantum-mechanical laws leads to violation of Bell's inequalities. If absolute trust is put in quantum mechanics, these inequalities, along with the assumption of 'local realism', should be discarded at once. However, local realism appears to be so natural and is so in line with our intuition that dedicated experiments were staged to verify Bell's inequalities.

The fulfillment of these inequalities have been verified (true, with polarized photons instead of 1/2 spin particles, but this is an equivalent situation) by various groups of experimenters. The first report was published by Aspect et al. [Aspect et al. (1981); Aspect (1982)]. It turned out that Bell's inequalities were violated. Consequently, local realism, or the assumption of an a priori existence of a distribution over spin projections (from which Bell's inequalities are derived) are experimentally refuted.

Positive results of the experiments of the type of Aspect means that the reality we meet in our world is not the same simple and intuitively clear concept of reality that is accepted in classical physics. Actual is what can be called *quantum reality*. One of signs of it is *quantum non-locality* revealed itself in the Bell's theorem and in the experiments of the type of EPR. Quantum reality reveals in special features of quantum measurements that are hard for understanding because of classical character of our intuition. One of the most transparent demonstrations of quantum reality is given

---

[8] "Realism" because a definite distribution is supposed to really exist even before they are measured, "local" because the measurement of one of the particles is supposed to be independent of the measurement of the other particle, located in another point.

by the so-called quantum games that will be shortly considered in Section 3.4.2.

Let us discuss in more detail the consequences of the Bell's theorem and Aspect's experiments for quantum mechanics and the 'measurement problem'.

This implies that the usual (and indispensable for classical physics) notion that the properties observed in a measurement actually exist even prior to the measurement, and that the measurement merely eliminates our lack of knowledge as to what specific property exists, turns out to be incorrect. In quantum measurements (i.e., for sufficiently precise measurements of quantum systems), this is not the case: the properties revealed in the measurement may not have existed prior to the measurement.

To explain this, we address ourselves again to the simple formulas given above. We consider a measurement that ascertains in which of the two states, $|\psi_1\rangle$ or $|\psi_2\rangle$, the system is (to put it differently, which of the two properties, numbered 1 and 2, the system has). The measurement gives a definite answer to this question, i.e., a choice is effected between the numbers 1 and 2, and after the measurement, the system does find itself in the state ($|\psi_1\rangle$ or $|\psi_2\rangle$) corresponding to the number chosen. Thus, the property indicated by the measurement result is inherent in the system after the measurement.

But did the system have this property prior to the measurement, i.e., was it in the state $|\psi_1\rangle$ or in the state $|\psi_2\rangle$ even prior to the measurement? Not at all. In general, prior to the measurement the system was in the state $|\psi\rangle = c_1|\psi_1\rangle + c_2|\psi_2\rangle$, which in general case is not identical to either $|\psi_1\rangle$ or $|\psi_2\rangle$.

The property exhibited in the measurement had not existed prior to the measurement. The apprehension of reality customary for classical physics, which is cognized in measurements, does not take place in quantum physics. In a sense, in a quantum measurement, *the reality is created and not merely cognized!* In point of fact, this implies that the classical apprehension of reality is never correct whatsoever, although in some cases, in relatively rough measurements, the classical perception of reality does not entail crude errors, i.e., provides a rather good approximation.

And now we have to clarify the statements made just above: precise formulations are needed in the problem under discussion, and the simple formulations that we have employed contain an inaccuracy.

We have said that the measurement exhibits some property and that the system indeed has this property after the measurement (although the system had not had it prior to the measurement). In terms of formulas, after the measurement that differentiates the states $|\psi_1\rangle$ and $|\psi_2\rangle$, the system does occur in one of these states. Is this indeed the case? No, undoubtedly we can make a somewhat weaker assertion: our consciousness tells us that the system occurs in either the state $|\psi_1\rangle$ or the state $|\psi_2\rangle$. Thus speaks our consciousness (such is the subjective impression), but whether this is so in reality (objectively) is a separate question.

If that which our consciousness tells us does (objectively) take place, we can formulate the following: if the measurement result is perceived by the observer, this ensures that the system is in one of the states $|\psi_1\rangle$ or $|\psi_2\rangle$. However, this is impossible to prove. Only a weaker statement is proved experimentally (we draw attention to how subtle the difference is): if the measurement result is perceived by the observer, the assumption that the system is in one of the states $|\psi_1\rangle$ or $|\psi_2\rangle$ will never lead to a contradiction with any further observations performed by this or any other observer.

The previous statement referred to the situation when the observer perceived the measurement result (looked and the measuring device). But if the observer does not look at the measuring device, the picture is different, even after the device was actuated. Then, the state of the combined system (the system being measured + the measuring device) is described by the vector

$$|\Psi\rangle = c_1|\psi_1\rangle|\Phi_1\rangle + c_2|\psi_2\rangle|\Phi_2\rangle.$$

This signifies that neither the system being measured nor the device reside in any definite (pure) state. Instead, the combined system they make up is in an entangled (quantum-correlated) state.

The chain of reasoning has now become so complicated that there is good reason to emphasize the central points. For us, the central point is the fact that the superposition that existed prior to the measurement does not disappear by the action of the device, at least until the observer becomes aware of the measurement result. After the measurement, the superposition $|\psi\rangle|\Phi_0\rangle = (c_1|\psi_1\rangle + c_2|\psi_2\rangle)|\Phi_0\rangle$ passes into the superposition $|\Psi\rangle = c_1|\psi_1\rangle|\Phi_1\rangle + c_2|\psi_2\rangle|\Phi_2\rangle$ and not into one of the factorized states that are the components of this superposition.

This is how it should be. It should be so because the quantum-mechanical evolution law is linear, it is described by the linear evolution operator or the linear Schrödinger equation. This law does not allow the sudden disappearance of all but one term of the superposition, as is implied by the picture of reduction occurring in the measurement. The state being a superposition of the vectors $|\psi_1\rangle|\Phi_1\rangle$ and $|\psi_2\rangle|\Phi_2\rangle$ cannot transform to one of this vectors.[9]

However, we immediately recall that the conversion of the superposition into one of the component of this superposition is just that transformation of *reduction*, which is involved in the ordinary, naive picture of a quantum measurement (see Sect. 3.3.1). The observer always subjectively perceives either $|\psi_1\rangle|\Phi_1\rangle$ or $|\psi_2\rangle|\Phi_2\rangle$. He always sees that only one component of the superposition persists. And because this always corresponds to observations, the change whereby all but one term of the superposition vanish was introduced into quantum mechanics by von Neumann's reduction postulate. The corresponding transformation is referred to as the state reduction, or von Neumann's projection, or the collapse of a wave function.

The simple description of quantum measurements based on the picture of reduction never leads to any errors in calculations of probabilities of the quantum effects. This description, and the picture of the state reduction during quantum measurements, are appropriate from the viewpoint of mathematical formalism and receipts of calculations in quantum mechanics. But we have seen that this description is incompatible with the linearity of quantum mechanics. This points onto a paradox, or internal conceptual problem of the theory. How can it be resolved?

Beginning with the early years of quantum mechanics, it was assumed that quantum-mechanical systems may evolve in two qualitatively different ways: as long as they are not measured they evolve linearly, and they undergo reduction in a measurement.

This postulate, adopted in the prevailing *Copenhagen interpretation* of quantum mechanics, has always worked perfectly, and continues to work just as remarkably nowadays. From the standpoint of practical needs, techniques of calculation, and predictions, there is no reason to abandon this postulate. Moreover, for practical computational needs, this postulate (and, of course, its different purely technical elaborations and generalizations) should undoubtedly be retained. But from what standpoint can it be doubted? Because it leads to correct predictions, is it not the proof of its correctness? There seems to be no other criterion in physics.

---

[9]In the most general case of linear transformation one of the components of a superposition can be transformed to zero vector. In quantum-mechanical evolution this is impossible because this evolution is not only linear but also unitary. However, in the specific situation of (ideal) measurement the condition of unitarity may be omitted. The only condition of linearity is enough to prevent disappearance of any component of superposition corresponding to one of the pointer basis vectors.

Yes, this is so. Those who make attempts to replace the reduction postulate with something qualitatively different do not have firm footing. And still there are grounds to make these attempts. We list now these grounds that are inside quantum mechanics. They are nevertheless no proof. Abandoning the reduction postulate would be justified only if the replacing theory is confirmed by practice in some way or another. We shall point out such more essential arguments later, in the subsequent chapters, but these arguments will be connected with the phenomenon of consciousness, i.e. something outside quantum mechanics (understood as theory of inanimate matter).

Now, what in the very quantum mechanics, in its conventional formulation, points to the necessity to reject the reduction postulate?

First, the search for another way that does not rely on the reduction picture is being continued in the attempts to eliminate the paradoxicality of quantum mechanics. A very promising avenue involves abandoning the reduction postulate in the framework of Everett's concept, which is discussed below. Second, the reduction postulate itself can be criticized. We briefly consider this criticism.

The reduction postulate appears to be alien to quantum mechanics and makes it *eclectic*. Why should a system evolve differently when it is subjected to measurement? Measurement is nothing more nor less than the interaction with some other system, conventionally termed the measuring device. Therefore, the evolution of the combined system (the measured one plus measuring device) during this interaction (i.e., during measurement) should be linear. The superposition does not disappear in the course of this evolution, and all components of the superposition that existed prior to the measurement persist after it as well.

It is significant, of course, that the measuring system is macroscopic, and hence the classical description is a good approximation for it. However, if this is merely an approximation, the exact, i.e., quantum-mechanical, description is equally applicable. After all, the measuring system consists of the same microscopic atoms, although in great number. That is why the conclusion that the superposition cannot vanish, reached in the framework of the quantum description, as well as its further implications, is not refuted by the fact that the measuring device is macroscopic.

Apart from the macroscopic nature of the device, also of significance is the fact that instabilities may emerge in the course of measurement to effectively lead to a situation resembling reduction. However, the 'derivation' of reduction with the aid of that kind of reasoning (see, e.g., Section 2.3 of

Ref. [Chernavskii (2001)]) also involves approximations. That is why it cannot refute the results of the analysis based only on one circumstance—the linearity of quantum mechanics, i.e., precisely the theory that was the starting point for these approximations.

Along the line of reasoning that we pursue in the subsequent discussion, the emphasis is placed on precisely the general properties of quantum mechanics. The purpose is to endeavor by analyzing these general properties (in the present instance, primarily linearity) to derive as much as possible for the understanding of the foundations of the theory and its interpretation. On this path, one has to make steps that sometimes look like fantasy. Such is the Everett's interpretation of quantum mechanics with its assumption of parallel worlds.

In our view, we may live in reconciliation with such steps to the extent to which they not only solve the originally formulated problem (overcoming the paradoxicality of quantum mechanics), but substantially broaden the area of application and the capabilities of the whole theory as well. However, just this will be achieved (in the following chapters) in the course of consideration and further extension of the Everett's many-world interpretation.

3.4.2 *Quantum games*

Quantum correlations considered above in connection with the notion of quantum reality are also seriously exploited in various quantum-informatical devices (among them quantum computers and quantum cryptographic schemes). However, unexpectedly quantum reality is demonstrated in so-called *quantum games*. Difference of quantum reality from classical concept of reality is quite transparently seen therein.

The quantum games are intellectual games in which some questions are asked and answered. Each of the players forming a command have to ask the questions offered to them (choose one of the set of allowed answers). The rules of the games are such that guaranteed victory of the command of players seems evidently impossible. At first sight the absolute impossibility of the guaranteed victory may be rigorously proved. Nevertheless, it turns out that the command of players may win with guarantee if they make use of special quantum devices. The solution of the paradox is that the "proof" does not take into account that reality of our world is quantum rather than classical. The special quantum devices realize advantage of the quantum reality.

The discrepancy between what seems evident (and even seems proved) and what is actually valid demonstrates in a clear way the classical character of our everyday intuition and quantum character of the reality in our world. The essence of these games may be expressed by the question: "How to win in the game if it is impossible to win in it?"

Let us begin with a simple example that is not astonishing if deception may be supposed in the game. Let the player created EPR-pair as is explained in Sect. 3.4.1, took one of the particles for himself and handed another one to his opponent. After this he said: please measure the spin projection of your particle onto the axis z, and I shall guess what will be the result of your measurement. It is evident that he shall be able to precisely guess the result of the measurement performed by this opponent. For this aim he has to measure the spin projection of his own particle. The result of the opponent will be opposite (because of the anticorrelation between the spin projection of the two correlated particles). The reason of necessary coincidence is (anti)correlation of the spin projections in the EPR-pair.

If one does not know about quantum aspects of this game but is sure that there is no deception in it, then he will be astonished by the guaranteed correct guess of the player.

More complicated games are constructed in such a way that circumvention is excluded by the very roles of the game. Here is one of them, proposed by the physicists D. M. Greenberger, M. A. Horne, and A. Zeilinger.

The game is arranged with the command of three players A, B and C in the following way:

- Any preparations are permitted before the game, but after this the players A, B and C are isolated from each other absolutely so that no communications are impossible between them during the game. For example any signals between them are absolutely excluded during time of the game Δt if the distance between any pair of the players is more that the time necessary for light issued by one of them to achieve the other ($l > c\Delta t$, where c is the light velocity).
- At a certain moment t each player is asked one of the two possible questions: "What is X?" or "What is Y?"
- Each player should answer $+1$ or -1.
- The answers of all players have to be given while the players cannot communicate.

The conditions of the game are as follows:

- Either all the players are asked about X, or one of them is asked about X, while two others are asked about Y.
- If all three players are asked about X, then the command of these players win in case if the product of their answers is equal to -1. Ir one of the players is asked about X, while two others are asked about Y, then the command wins if the product of their answers is equal to $+1$.

Let us consider how the command of the three players can win with guarantee. It seems evident that this is impossible. Indeed,

- Since any communication during the time of the game is excluded, the answers of the players to all possible questions may be prepared beforehand. Denote these prepared numbers $\{X_A, Y_A; \ X_B, Y_B; \ X_C, Y_C\}$, each of them equal to $+1$ or -1 (here X_i is the answer of the ith player to the question about X, and Y_i is the answer of the ith player to the question about Y).
- For the guaranteed victory, these numbers have to satisfy the following equations:

$$X_A X_B X_C = -1,$$
$$X_A Y_B Y_C = 1, \quad Y_A X_B Y_C = 1, \quad Y_A Y_B X_C = 1.$$

- These equations are incompatible (product of all left-hand-sides of them is a full square i.e. positive, while the product of all right-hand-sides is -1.

Despite of this seemingly correct proof of impossibility, the guaranteed victory in this game is possible. The "proof" does not take into account existing of quantum-correlated systems. The solution leading to the guaranteed victory is in exploiting by the players three spin $1/2$ particles in the special correlated state:

$$|GHZ\rangle = \frac{1}{\sqrt{2}}\left(|z^+\rangle_A |z^+\rangle_B |z^+\rangle_C - |z^-\rangle_A |z^-\rangle_B |z^-\rangle_C\right)$$

where $z^\pm$ denotes the state of the corresponding spin with the projection $s_z = \pm 1/2$.

After preparation of this state of the three spin $1/2$ particles they are given to the three players. In the game, each of them before answering a question (about X or about Y) measures the corresponding projection of the spin in his possession (s_x if the question of X is asked, s_y in case of the question about Y is asked). The answer should be $+1$ if the corresponding projection is positive and -1 in case of the negative projection.

One can prove that the spin projections obtained in the measurements satisfy the following equations equivalent in fact to the above equations required for the victory (the notations $\sigma_x = 2s_x$, $\sigma_y = 2s_y$ are introduced here):

$$\sigma_{xA}\sigma_{xB}\sigma_{xC} = -1,$$

$$\sigma_{xA}\sigma_{yB}\sigma_{yC} = 1, \quad \sigma_{yA}\sigma_{xB}\sigma_{yC} = 1, \quad \sigma_{yA}\sigma_{yB}\sigma_{xC} = 1.$$

These equations are easily derived if the correlated state $|GHZ\rangle$ is expressed in terms of $x^{\pm}$ and $y^{\pm}$ instead of $z^{\pm}$. For example, the transition to $x^{\pm}$ is performed by the following relations:

$$|z^{+}\rangle = \frac{1}{\sqrt{2}}\left(|x^{+}\rangle + |x^{-}\rangle\right), \quad |z^{-}\rangle = \frac{1}{\sqrt{2}}\left(|x^{+}\rangle - |x^{-}\rangle\right).$$

The resulting form of the state $|GHZ\rangle$ is

$$|GHZ\rangle =$$
$$\left(|x^{+}\rangle_A + |x^{-}\rangle_A\right)\left(|x^{+}\rangle_B + |x^{-}\rangle_B\right)\left(|x^{+}\rangle_C + |x^{-}\rangle_C\right)$$
$$- \left(|x^{+}\rangle_A - |x^{-}\rangle_A\right)\left(|x^{+}\rangle_B - |x^{-}\rangle_B\right)\left(|x^{+}\rangle_C - |x^{-}\rangle_C\right)$$

or, after some algebra,

$$2|GHZ\rangle =$$
$$|x^{-}\rangle_A|x^{-}\rangle_B|x^{-}\rangle_C$$
$$+ |x^{+}\rangle_A|x^{+}\rangle_B|x^{-}\rangle_C + |x^{+}\rangle_A|x^{-}\rangle_B|x^{+}\rangle_C + |x^{-}\rangle_A|x^{+}\rangle_B|x^{+}\rangle_C.$$

It shows straightforwardly that the result of the measurements by each players the x-th projection of his spin will give $\sigma_{xA}\sigma_{xB}\sigma_{xC} = -1$.

Analogously, other forms of the state $|GHZ\rangle$ (that are necessary to prove other equations guaranteeing the victory) may be derived with the help of the transformation formulas

$$|z^{+}\rangle = \frac{1}{\sqrt{2}}\left(|y^{+}\rangle - i|y^{-}\rangle\right), \quad |z^{-}\rangle = -\frac{i}{\sqrt{2}}\left(|y^{+}\rangle + i|y^{-}\rangle\right).$$

Thus, we see with great evidence that the arguments based on the purely classical concept of reality (we can prepare the answers before the game because no communication is possible during the game) lead to erroneous conclusions. In our world quantum reality is valid, and according to it the measurement results emerge in the process of measurement rather than are predetermined before it. Accounting quantum character of the reality implies sometimes counter-intuitive but nevertheless always correct conclusions. The advantage arising because of the quantum character of reality is exploited in the so-called quantum information technologies to provide for unexpected but quite real new capabilities (for example transferring secret messages with absolute security).

3.4.3 *Quantum reality from various viewpoints*

Let us briefly consider two papers published in Physics–Uspekhi in the course of the discussion on the 'measurement problem'.

(1) A V Belinskii's paper [Belinskii (2003)] contains an interesting technical remark about Bell's inequalities with the inclusion of detector errors. The author concludes that Aspect-type experiments, despite the finite detector accuracy in these experiments, reliably refute Bell's inequalities even without further improvement of the detectors, thereby experimentally bearing out the nonexistence of local realism in nature.

But the bulk of the contents of Ref. [Belinskii (2003)] is concerned with another issue. By the thoroughly analyzed specific examples of real or thought experiments with photons, Belinskii illustrates the major distinction of quantum measurements that generates the 'measurement problem': the property of a system revealed in its measurement (for instance, a specific photon polarization) might not have existed prior to the measurement. This proposition of the quantum theory of measurement, which is central to the 'measurement problem', was analyzed in detail in the foregoing. However, in the examples given by Belinskii, it appears in a light that is supposedly more convincing for those physicists who are used to dealing with descriptions of specific experimental facilities rather than with abstract reasoning.

For instance, Belinskii considers an experiment in which single photons are detected (emitted by a source so low in intensity that the probability of simultaneous arrival of more that one photon at the detector is negligible). In this case, evidently, it is possible to count the number of detector actuations and thereby find the number of arriving photons. All this appears so evident that we do not notice when we involuntarily yield to the temptation to use the intuition borrowed from classical physics.

But Belinskii puts forth questions that do not permit one to lapse into thinking thus: "It is commonly believed that photo-counts, or bursts of the detector's photo-current, correspond to the arrival of photons. But is it so? Do quanta really exist in the light field? The detector measures the number of photons in the field. But does a definite value of this quantity exist before the measurement?" And it turns out that simple experiments can prove that the answers to these questions are negative: the light field cannot be represented as an ensemble of a definite number of photons, the number of photons is not defined prior to the instant of measurement.

For instance, the photon source can be made such that one photon is recorded at times and two photons at other times. The field should seemingly consist of single photons and photon pairs. However, this is not so, which can be proved experimentally. It would be inappropriate to go into details here. The interested reader is referred to Ref. [Belinskii (2003)], where the logic of experimenters who refute the classical notion of the number of photons is traced in detail.

(2) To analyze the operation of the observer's consciousness, A. D. Panov [Panov (2001)] makes use of the notion of decoherence, which is undeniably of paramount importance in this context. Panov discusses the decoherence occurring in a material substance, which is responsible for the realization of the measurement result by the observer (e.g., in a special material structure in the brain). The endeavor to reduce everything to ordinary physical processes occurring in physical systems is quite natural for physicists and has always constituted one of the main areas of work on the problem. And the physically clear decoherence effect is undoubtedly an appropriate instrument for endeavoring to realize suchlike reduction.

Panov makes a very important observation that the entanglement of two quantum systems (in the present instance, the system being measured and the material substance in which the measurement result is reflected, or perceived) leads to the decoherence of both of them. When the density matrix of the system being measured contains, after the interaction with the device, components corresponding to all measurement results, the same statement is applied to the density matrix of the observer's brain.[10]

Let us elucidate this statement. Consider the previously introduced states $|\Psi_1\rangle = |\psi_1\rangle|\Phi_1\rangle$ and $|\Psi_2\rangle = |\psi_2\rangle|\Phi_2\rangle$ of the system being measured and the device, which correspond to definite measurement results. As we have seen, in reality, the state $c_1|\Psi_1\rangle + c_2|\Psi_2\rangle$ sets in after the measurement. We now include the observer (the observer's body or, say, the observer's brain) into the description. When the measurement has been effected but the observer has not yet become aware of the result (for instance, has not looked at the scale of the device), the combined state of the system being

---

[10]Panov tells of the density matrix of "the observer's consciousness", but actually he means the state of some material structure in the body of the observer, for example in his brain.

measured, the device, and the observer is given by $(c_1|\Psi_1\rangle + c_2|\Psi_2\rangle)|\chi_o\rangle$ But once the observer has realized the measurement result (for example, the photons emitted by the device arrive at his eye and his brain properly reacts to this signal), the state becomes

$$|\Omega\rangle = c_1|\Psi_1\rangle|\chi_1\rangle + c_2|\Psi_2\rangle|\chi_2\rangle,$$

i.e., the entanglement of the system being measured with the device and the observer occurs.

Then, the system under measurement and the device, which are considered separately from the observer, cannot be characterized by a definite state vector. Instead, the system under measurement combined with the device (but without the observer) can be characterized by the density matrix:

$$\rho_\Psi = \mathrm{Tr}_\chi\left(|\Omega\rangle\langle\Omega|\right) = |c_1|^2|\Psi_1\rangle\langle\Psi_1| + |c_2|^2|\Psi_2\rangle\langle\Psi_2|.$$

The density matrix describes not a pure state but a mixed state of the system under measurement and the device considered as a unified system. The density matrix signifies that this system is in the pure state $|\Psi_1\rangle$ with the probability $|c_1|^2$ and in the pure state $|\Psi_2\rangle$ with the probability $|c_2|^2$. In other words, decohering of the system comprising the system under measurement and the device occurred, brought about by the interaction of this system with the observer.

It is significant, however, that the observer's state also underwent decohering in this case.

Indeed, proceeding from the entangled state $|\Omega\rangle$ and trying to describe the state of only the observer himself, we can achieve this by applying the procedure of taking the partial trace again, but this time the trace should be taken over all systems except the observer himself. For the observer (considered as a physical system), we then obtain the density matrix

$$\rho_\chi = \mathrm{Tr}_\Psi\left(|\Omega\rangle\langle\Omega|\right) = |c_1|^2|\chi_1\rangle\langle\chi_1| + |c_2|^2|\chi_2\rangle\langle\chi_2|.$$

The mixed state of the observer represented by this density matrix is interpreted in an obvious way: it is in one of the pure states $|\chi_1\rangle$ and $|\chi_2\rangle$ with the probabilities $|c_1|^2$ and $|c_2|^2$. It is significant that the mixed states both of the observer and of the system under measurement and the device are characterized by the same probability distribution.

Although we have been speaking of the observer's state for simplicity, in reality we are dealing with some material carrier of the observer's consciousness (e.g., with some structure in his brain). We see that when considering this structure, we obtain the decoherence picture that perfectly corresponds to the decoherence of material systems outside the observer.

Such an analysis is undoubtedly beneficial for the understanding of what takes place. But does it solve the measurement problem?

It is evident that in the description of a measurement, we cannot restrict ourselves to only the description of the observer's decoherence but endeavor to make one more step and pose the following question: what in reality occurs after the measurement? Does the observer remain in one of the pure states $|\chi_1\rangle$, $|\chi_2\rangle$ after the measurement or should we think, being guided by the form of the state vector $|\Omega\rangle$, that none of these states can disappear and they all persist as the components of the superposition $|\Omega\rangle$? If we opt for the latter, we once again encounter a paradoxical situation and the 'measurement problem': quantum mechanics compels us to believe that both states $|\chi_1\rangle$ and $|\chi_2\rangle$ continue to exist (in the superposition), while 'worldly wisdom' shows that the observer always 'perceives' only one of them.

Sometimes one encounters the opinion that the effect of decoherence eliminates the 'measurement problem'. The arguments may be the same as in the Panov's paper. We nevertheless, even with the reasoning of this type taken into account, adhere to the standpoint that decoherence, while significantly elucidating the situation with quantum measurements, does not remove all the questions that have led to the 'measurement problem'. To advance further, the analysis should, in our opinion, be continued. Following this logic, we revert to the discussion of the role of consciousness in quantum measurements.

3.5   Measurement problem: stages of investigation

3.5.1   *Formulation of the problem*

The problem that we are trying to outline is often referred to as the 'measurement problem'. It was posed at the dawn of quantum mechanics and reflected the aspiration of moving beyond the framework of the Copenhagen interpretation (associated primarily with Bohr's name), which perfectly solved practical problems but left some discontent from the conceptual standpoint. Attempts to solve the measurement problem were made by many outstanding physicists, including Pauli, Schrödinger, Heisenberg, and Einstein (and, of course, Bohr himself with his brilliant analysis of the special features of quantum mechanics). However, even today, this problem is not considered solved.[11]

It is not so easy to trace tendencies in the attitude of the physical community toward the 'measurement problem', because every generation of physicists begins to comprehend it to some extent anew and is able to introduce something new in its solution only after an arduous and long period of familiarization with the problem. Nevertheless, it seems to us, we can distinguish three qualitatively different stages in the investigation of this problem.

3.5.2 *Enthusiasm and optimism*

The first stage, when all the founding fathers of quantum physics addressed this subject to some extent, was noted for enthusiasm and optimism of researchers. The enthusiasm and interest were maintained by the fact that the problem ushered physicists into an entirely new, previously unknown and therefore interesting realm of meta-science and philosophy, leading them to compare the existing and newly emerging specific propositions of science with the most general methodological issues and quite frequently with world outlook. The optimism, which is quite natural at the inception, also generated because extremely potent intellectuals participating in the research.

At that period, different lines were explored. But serious advances were made only along one of them: the Copenhagen interpretation of quantum mechanics, which relied on the von Neumann reduction postulate, was formulated and polished to the state of a clear algorithm. In point of fact, this interpretation was a compromise, which made it possible to work in quantum mechanics having no doubt as to the correctness of this work. In essence, the conceptual difficulties were not overcome, but those who were not concerned with them could forget about them without the apprehension of losing orientation in practical quantum-mechanical calculations.

3.5.3 *Marginalization*

The second stage began when it became clear that the first results contributed little to the solving of the 'measurement problem' except maybe a better understanding of the problem itself, its extraordinary nature, and its scale.

---

[11]We think that this problem is actually solved by the Everett's interpretation (see below), but this opinion is not commonly accepted.

This stage was characterized by a nearly universal belief in the Copenhagen interpretation resulting in the marginalization of the 'measurement problem' with its requirement to go beyond this interpretation. The time had passed when the understanding of quantum mechanics (at the intuitive level) seemed to be, and indeed was, indispensable to efficient work. There now existed a clearly formulated system of rules, and obtaining results in the framework of this system required only the mathematical treatment of a specific problem, i.e., calculations. The issue of understanding came to seem superfluous, and the majority of physicists were no longer concerned with it.

Papers on the 'measurement problem', which would nevertheless appear from time to time, changed in character and became more scholastic. Proposed instead of bold new solutions were different formulations of the old ones, which changed these old formulations in so subtle a verbal nuance that the significance of changes was clear (and interesting) to only a narrow circle of active participants in the discussion. The majority of physicists considered this discussion wholly irrelevant to physics.

3.5.4 *Everett's "Many-Worlds" interpretation*

In 1957, young American physicist Hugh Everett [Everett (1957)] came up with a very bold and radically new 'interpretation of quantum mechanics that was called later Many-Worlds' interpretation. This marked the beginning of a new stage in the investigation into the 'measurement problem'. Everett's paper was initially noticed by few. Such famous physicists as Brice DeWitt and John Wheeler [DeWitt and Graham (1973)] were among those who became interested in this paper, which nevertheless remained unnoticed by the broad scientific community. However, it played and continues to play the leading part at the new stage of investigation.

This stage properly commenced approximately two decades ago and continues to the present day. The interest in the 'measurement problem' has remarkably quickened and the people engaged in the problem significantly grew in number. There were reasons for these changes. Quantum mechanics had essentially changed to become an engineering science, and therefore the overall number of physicists involved in it became much greater than before. Furthermore, all the preceding development of quantum mechanics had shown that it can find application in quite unexpected areas, and hence the quest for and mastering of new applications to an increasing extent called for people unconstrained by tenets. All this changed the very atmosphere of the quantum-mechanical community and significantly moderated its conservatism.

There were also more specific reasons for rekindling the interest in the conceptual problems of quantum mechanics, in the 'measurement problem'. Required were not only calculations of the ensembles of quantum systems (atoms, electrons, photons, etc.), but of individual systems as well (a single electron in single-electron devices, a single ion in a magnetic trap, etc.). The 'ensemble' ideology was no longer quite suited to describe the behavior of suchlike systems. It was necessary to be able to describe not only an ensemble of systems but also an individual system. Furthermore, for purely practical purposes (e.g., in quantum optics), the demand existed to calculate not a single measurement but a series of measurements performed over the same individual system or a measurement continuous in time. In these conditions, the statement insistently repeated in textbooks on quantum mechanics that the state vector (wave function) describes a quantum ensemble rather than an individual system came to generate increasingly more discontent. The ensemble ideology, in which there emerge no conceptual problems at all, became manifestly insufficient.

Furthermore, there appeared qualitatively new applications of quantum mechanics whose realization required a far deeper understanding of the specific character of quantum mechanics. These new applications were united under the common title *quantum informatics* to embrace quantum cryptography, quantum teleportation, and, above all, quantum computing. The new technologies that emerged on this basis employed precisely those specific features of quantum systems which generate the 'measurement problem'. The development of quantum-information systems in general and quantum computers in particular invited a considerably deeper understanding of the essence of quantum mechanics and its distinctions from the classical one. In addition, it was necessary to be able to correctly describe the behavior of such systems, which have quantum and classical properties simultaneously.

Of course, it is invalid to say that solving the 'measurement problem' was required before solving practical technological problems. However, developing methods for solving practical problems invited work at an extremely high level of understanding of quantum mechanics, which is close to the level of formulation of this problem. This broadened the circle of those concerned with the conceptual problems of quantum mechanics and the circle of those who worked actively in this area.

3.6   "Many-Worlds" interpretation and separation of alternatives

What direction does the quest for solving the conceptual problems take? Not pretending to present complete coverage, we mention only one direction, which is supposedly the principal one. This is a return to Everett's concept (or interpretation) [Everett (1957)] often known as the "Many-Worlds" interpretation of quantum mechanics.

3.6.1   Relative states

Everett himself called it *the relative state interpretation* of quantum mechanics; however, more recently, after Wheeler's and DeWitt's papers [DeWitt and Graham (1973)], it came to be known as the Many-Worlds interpretation. This name owes its origin to the fact that Everett's concept permits the existence of numerous (actually, an infinite number) of classical realities, which may be intuitively represented as the set of classical worlds. The Many-Worlds, or Everett's, interpretation, which was earlier considered too fantastic, has been actively discussed and adopted by many scientists. Many aspects of this interpretation were thoroughly studied and different versions of its development were proposed (see [Vaidman (2002)] for a rather comprehensive review of the literature on this subject).

In what is following, far from being a complete reflection of all viewpoints, we highlight only some minimal and yet logically complete lines of reasoning, which have, in our opinion, attractive new prospects.

We first of all explain Everett's interpretation (concept) by continuing the logic of reasoning started in the preceding sections.

We adduced plausible reasoning testifying to the fact that von Neumann's reduction postulate is alien to quantum mechanics and has been adopted in it (at the cost of eclecticism) only to evade conceptual problems rapidly and easily, not solving them in essence, and go over to practical calculations.

In the case of von Neumann's reduction, of the initial superposition in the previously used example $c_1|\Psi_1\rangle + c_2|\Psi_2\rangle$, there remains only one component (e.g., $|\Psi_1\rangle$ or maybe $|\Psi_2\rangle$). But at variance with this picture, the linearity of quantum mechanics requires that all terms of the superposition should persist. In the measurement, there only occurs entanglement of the system under measurement and the environment, i.e., the superposition takes the form $c_1|\psi_1\rangle|\Phi_1\rangle + c_2|\psi_2\rangle|\Phi_2\rangle$.

Everett's concept may be treated as an attempt to seriously make this argument and consistently take it into consideration.

Following the Everett's logic, We therefore attempt to be consistent and not 'spoil' quantum mechanics by its alien reduction postulate but, conversely, rely on its immanent linearity. We are then forced to conclude that after the interaction, which we term the measurement, the state of the system and the device assumes the form $c_1|\psi_1\rangle|\Phi_1\rangle + c_2|\psi_2\rangle|\Phi_2\rangle$. None of the components of this superposition may be discarded, contrary to the Copenhagen interpretation with the von Neumann's postulate. All the components of the superposition have to be dealt on equal footing, all of them have to be considered "equally real".[12]

If these superposition terms are not discarded, they all are to be interpreted. This is precisely what Everett did. In Everett's concept (more precisely, in the equivalent Many-Worlds interpretation), different terms of the superposition are assumed to correspond to different classical realities, or classical worlds. These realities, or worlds, are assumed to be exactly equivalent, i.e., none of them is more real than the others. As a result, we obtain the Many-Worlds picture in the Everett–Wheeler–DeWitt sense.

This is one of the formulations of *quantum reality*. Various terms of the superposition presents various classical realities that are incompatible with each other, alternative to each other. However, only all of these alternative classical realities (in a simpler wording, *classical alternatives*) presents what may be called *quantum reality*.

This line of argument has to be compared and somehow agreed with the fact that any experimenter observes only one (of all possible) measurement results, so that his experience seems contradictory to the treatment of all results (all "classical realities") on equal footing. In this point essential becomes *consciousness of an observer*.

3.6.2 *Separation of the alternatives by consciousness*

The contradiction mentioned at the end of the preceding section concerns in fact not what exists but what an observer perceives. Therefore, it concerns of the observer's consciousness.

And what is to be done with the consciousness in the Many-Worlds interpretation of quantum mechanics? Because every observer sees only one measurement result out of two (or many). Is this at variance with the

---

[12]In general, the superposition may contain many or even an infinite number of components, depending on the type of the measurement.

Many-Worlds concept? The apparent contradiction is solved quite easily: it is as if the observer's consciousness splits (is divided) in the "components", one for each of the parallel (Everett's) worlds (alternative classical realities). Then in every one of the classical worlds the observer sees what takes place in this world. We now show this.

Let the vector $|\chi_0\rangle$ denote the initial state of the observer (his body, or his brain) when he has not yet become aware of the measurement results (maybe it has not yet been completed or maybe he has not yet looked at the devices). Let $|\chi_1\rangle$ (accordingly $|\chi_2\rangle$) denote his state at the moment when he already knows that the measurement yielded result 1 (accordingly 2). Then, the system of three (the system under measurement + device + observer) prior to the measurement is in the state $(c_1|\psi_1\rangle + c_2|\psi_2\rangle)|\Phi_0\rangle|\chi_0\rangle$, after the measurement but prior to perceiving the measurement result is in the state $(c_1|\psi_1\rangle|\Phi_1\rangle + c_2|\psi_2\rangle|\Phi_2\rangle)|\chi_0\rangle$, and after perceiving it in the state $c_1|\psi_1\rangle|\Phi_1\rangle|\chi_1\rangle + c_2|\psi_2\rangle|\Phi_2\rangle)|\chi_2\rangle$.

Everett's' interpretation of this expression is evident: in every one of the classical worlds, the observer sees (realizes) that which took place in precisely this world. In the world denoted by number 1, the observer is in the state $|\chi_1\rangle$. This signifies that he has perceived the measurement yielding result 1, i.e., that the system being measured and the device are in the state $\psi_1\rangle|\Phi_1\rangle$. Similarly, in the world number 2, the observer is in the state $|\chi_2\rangle$, i.e., in his consciousness, the picture of what is taking place corresponds to the state $\psi_2\rangle|\Phi_2\rangle$ of the system under measurement and the device (see Fig. 4.1 on page 78).

Therefore, *the observer's consciousness splits, is divided, in accordance with how the quantum world is divided into the ensemble of alternative classical worlds.* In our example, there are only two alternatives; generally, the number of alternative classical worlds turns out to be equal to the number of alternative results that the measurement may yield. We note, however, that in reality, the number of classical worlds may be arbitrarily large, even infinite, and after the measurement they split into classes (also infinite in this case) corresponding to alternative measurement results.

In the ordinary (Copenhagen) picture, a reduction of a state or, the equivalent, a selection of one alternative measurement result of all possible ones occurs. This may be termed *selection of an alternative*. All except the selected alternative vanish after the reduction. Going to Everett's interpretation, we see that no reduction, or selection, of a single alternative occurs. Instead, splitting, or division, of the quantum world state into alternative 'classical realities', or parallel worlds (Everett's worlds) occurs.

The observer's consciousness perceives different classical worlds independently of each other, or *consciousness separates the alternatives*. We can conventionally say that the consciousness splits into components, each of which perceives only one classical world. The observer subjectively perceives what is going on in such a way as if there exists only one classical world, specifically that which he sees around him. Instead, we can say that, according to Everett's concept, it is as if many 'replicas' of the observer exist, one replica for each of the parallel worlds. Sensations of each of these replicas provide to each of them the picture of precisely the world he 'lives' in.

In Everett's interpretation, there appears some duality, which is rather hard to comprehend. All alternatives are realized, and the observer's consciousness splits between all the alternatives. At the same time, the individual observer subjectively perceives what is going on in such a way as if there exists a single alternative, the one he exists in. In other words, the consciousness as a whole splits between the alternatives but the individual consciousness subjectively chooses (selects) one alternative.

To avoid misunderstanding, we note that in one (any) of Everett's worlds, all observers see the same thing, their observations are consistent with each other (unless, of course, we are dealing with possible purely human errors, but we assume perfect observers). This follows because, owing to the linearity of quantum-mechanical evolution, the initial state

$$(c_1|\psi_1\rangle + c_2|\psi_2\rangle)|\Phi_0\rangle|\chi_0^{(1)}\rangle|\chi_0^{(2)}\rangle$$

of the system being measured, the device, and two observers passes into the state

$$c_1|\psi_1\rangle|\Phi_1\rangle|\chi_1^{(1)}\rangle|\chi_1^{(2)}\rangle + c_2|\psi_2\rangle|\Phi_2\rangle)|\chi_2^{(1)}\rangle|\chi_2^{(2)}\rangle.$$

The states involving the factors $|\chi_1^{(1)}\rangle|\chi_2^{(2)}\rangle$ or $|\chi_2^{(1)}\rangle|\chi_1^{(2)}\rangle$, which would imply the inconsistency of observations, can in no way appear. Let us underlie that this is not a special assumption, but a simple mathematical feature following from the usual quantum-mechanical formalism.

3.6.3 *Discussion of the Everett's concept*

This is Everett's concept in brief. At first, it seems fantastic and too complicated. But this is not exactly so. First, Everett's concept logically follows from the single and seemingly quite natural assumption that the linearity of quantum mechanics is not violated in the course of interaction between the system under measurement and the measuring device and the subsequent action of the device on the observer.

Second, the entire picture seems more fantastic than it actually is when they speak, endeavoring to speak with clarity, about many classical worlds. In actual fact, not only does the Many-Worlds picture excessively dramatize the situation, but may also mislead (and quite often does so) those who familiarize themselves with it without sufficient background in this problem.

There is good reason to recall from time to time (and to necessarily do so whenever difficulties or hesitation show) that in reality, no 'many classical worlds' exist at all. There is only one world, and this is a quantum world, and it is in the superposition state. It is simply that every component of the superposition taken separately corresponds to what our consciousness perceives as the picture of the classical world, and to different superposition terms there correspond different pictures. What we call "classical (Everett's) world" is just one 'classical projection' of the quantum world. These different projections are produced by the observer's consciousness (perceived subjectively), while the quantum world itself exists independently of whatever observer (objectively).

When we say 'different superposition components' instead of 'different classical worlds', many misunderstandings that occur in the popular literature and in discussions on this issue disappear. For instance, in case of poor understanding of the Everett's concept, the Many-Worlds picture of measurement may create the illusion that one classical world transforms into several (or even an infinite number) of worlds at the instant of measurement. In this case, they sometimes even speak about a monstrous nonconservation of energy under this 'multiplication of worlds', or !world branching'.

In reality, there is of course nothing of the kind in Everett's interpretation. Prior to the measurement, as well as after it, there exists the single state vector that describes the state of the quantum world. At the instant of measurement (more precisely, at the instant of the interaction between the system being measured and the device), specific changes in this state and in its describing vector occur: the entanglement between the system being measured and the measuring device (the measuring medium).

For a formal description of this change, we represent the state vector as a superposition of several components and show how each of these components changes in the measurement (in the interaction). This analysis was discussed at length in the previous sections.

Let us make one more remark, concerning some technical details (not important in principle). Not only is the picture of world branching oversimplified, but so is the mere idea that the measurement takes place simultaneously at all points of a finite domain (in which the wave function of the system under measurement is nonzero) at a specific time instant. In particular, this is incompatible with the special theory of relativity, in which the simultaneousness of events at different points cannot be determined at all. All these difficulties arise from the idealization contained in the notion of instantaneous measurement. They disappear in going over to the picture of continuous measurement (in this connection, see Ref. [Mensky and von Borzeszkowski (1995)], where the measurement of position is discussed in the relativistic theory framework). Below, in Section 5.1, we discuss continuous measurement in greater detail and in this connection introduce another method of describing alternatives — with the aid of corridors of paths. With this description, the question of classical world 'multiplication' does not arise at all.

There is one truly significant objection to Everett's concept. It consists in the fact that this concept is impossible to verify, or at least it appears so at first glance. Because all formulas in it are the same as in the standard quantum mechanics, the predictions obtained in the framework of this concept are no different from those that follow from the standard quantum-mechanical calculations carried out in the framework of the Copenhagen interpretation. This is precisely the reason why Everett's concept is merely a different interpretation of quantum mechanics and not a different quantum mechanics.

Therefore, it appears at first sight that the Many-Worlds interpretation is impossible to confirm or refute by experiment, and in some sense this is so indeed. This is a serious drawback, because constructing a rather (conceptually) complex interpretation that is impossible to verify seems to be too high a price to be paid for making the theory more consistent in the purely logical aspect. This is the reason why several of Everett's proponents suggested that his concept should be modified so as to make it verifiable.

We believe, however, that Everett's concept can be verified, even without any modification, by resorting to experiments or, rather, to observations of a special kind, specifically, *observations of individual consciousness*. This is discussed later on, and we now try to specify more precisely how consciousness is to be treated in the framework of Everett's concept.

3.7   Conclusion: Subjective aspect in quantum mechanics

If we restrict ourselves to a very brief formulation, the immanent feature of quantum mechanics (more precisely, of quantum physics, including relativistic physics) that distinguishes it from all remaining physics is that attempts to represent the measurement process in it as completely objective, as absolutely independent of the observer who perceives the result of the measurement, have not met with success. To simplify matters still further, we say that the description of quantum measurements (at least if this is to be logically complete, consistent) must involve not only the system under measurement and the instrument but also the observer or, to be more precise, *the observer's consciousness*, in which the result of the measurement is fixed.

This feature of quantum mechanics contradicts our intuition and inevitably leads to misunderstanding upon first acquaintance. The complex of questions emerging in this connection is most frequently grouped under the conventional name 'measurement problem'. The several decades that have passed since the advent of quantum mechanics have shown that attempts to satisfactorily solve this problem or dismiss it as being nonscientific have been unsuccessful.

We think that the main reason of this failure is in the (explicit or implicit) conviction that the problem should be solved in the framework of completely objective form of science. We shall show in the subsequent chapters that the progress in solving the 'measurement problem' may be achieved if this arbitrary requirement is abandoned.

The 'measurement problem' concerning paradoxical features of quantum mechanics has quite special status. Some physicists are even lacking a clear understanding of the essence of this problem. This is partly attributable to the fact that the conceptual problems of quantum mechanics do not play a role of any significance in practical work on the calculation of quantum systems and are therefore uninteresting to physicists oriented to practical problems. This underlies the standpoint shared by many that the 'measurement problem' is far-fetched and scholastic, although the fact that this problem has been investigated by outstanding physicists can hardly permit disregarding it so easily.

We explained in this chapter, why the procedure of state reduction (collapse of the wave function), involved in the universally accepted description of a quantum measurement, is in essence a departure from quantum mechanics. Instead, it is possible to invoke the concept of 'observer's consciousness' by introducing it explicitly into the description of measurement. This is done in Everett's interpretation (Many-Worlds interpretation) [Everett (1957); Zurek (1998)].

In the next chapter, we shall go behind the Everett's interpretation, to Extended Everett's Concept (EEC), relying on the hypothesis about identification of the 'observer's consciousness' with the separation of the quantum world (in the subjective perceiving this *quantum* world) into classical alternatives corresponding to alternative results of measurements [Mensky (2000a)].

The special role of the 'observer's consciousness' underlies the Many-Worlds interpretation. Nevertheless, the complete identification of the consciousness with what takes place in the measurement leads to a radical change of the viewpoint on the problem as a whole and especially on the phenomenon of consciousness. As a result, there emerges a direct relation between physics and psychology and, from a more general standpoint, between the realms of human cognition represented by the sciences and the humanities, and more generally, with the spiritual sphere of knowledge.

Chapter 4

Consciousness in parallel worlds

"The general problem of the relation between psyche and physis, between inside and outside, can hardly be regarded as solved by the term 'psy-chophysical parallelism' advanced in the last century. Yet, perhaps, modern science has brought us closer to a more satisfying conception of this relationship, as it has established the notion of complementarity within physics. It would be most satisfactory if physis and psyche could be conceived as complementary aspects of the same reality."

W. Pauli

Letter by Pauli to Pais of August 17, 1950. Letter 1147 in von Meyenn (1996), p. 152. Cited according to [Atmanspacher and Primas (2006)].

Quantum mechanics as a special branch of physics that appeared at the beginning of 20th century and radically changed the views of scientists on what is reality. It turns out that this novel concept of reality is important not only for the internal needs of physics but also for much more general tasks of explaining the human experience.

Particularly, the concept of quantum reality can explain what is consciousness and why consciousness shows sometimes mystical features, particularly super-intuition and probabilistic miracles. We shall demonstrate this in the present chapter and elaborate the relevant ideas in the subsequent chapters.[1]

---

[1]The idea of Pauli in Jung about direct connection of quantum mechanics with consciousness originated from the analysis of *"sinchronisms"*. Jung used this term for the strange, inexplicable coincidences that sometimes happen. In the context of our approach this phenomenon may be interpreted as a sort of probabilistic miracles.

As a starting point for our analysis we need the concept of *quantum reality*. We shall formulate it here in a simple way as the *coexisting parallel classical realities* (classical alternatives), or, equivalently, as *parallel worlds*. Instead of the term "Universe", usually applied for our world (considered to be classical), we may apply the term *"Alterverse"* for the world in which quantum reality (presented as the superposition of the parallel classical worlds) rules.[2] It is clear that the subjective experience of living in Alterverse may be quite different from that in the Universe. The topic of the present book may be formulated as an attempt to answer the question: "How life in Alterverse differs from that in Universe?"

For those readers who are familiar with quantum physics this concept has been presented in more details in the preceding chapter, but these details are not necessary for reading the present and following chapters.

4.1   Parallel worlds (classical alternatives) as quantum reality

One of the main (actually the most important) differences of quantum mechanics as compared with the classical physics is that it admits *superposition of states*. For example, a point-like particle, both in classical and in quantum mechanics, can be localized in a point A or in a point B. However, in quantum mechanics the superposition of these two localized states of the point-like particle is possible. If the particle is in the state of superposition, $\psi = \psi_A + \psi_B$, one cannot say in what of these two points it is localized. And this is not because of the lack of knowledge since the state of the particle ψ is known.

Let us clarify the last statement. In classical physics we often meet the situation when it is not known in what point the given point-like particle is localized. However, this is only the consequence of the lack of knowledge: the particle is of course localized in some point (because it is point-like) but we do not know what point precisely. The situation in quantum mechanics is qualitatively different: the state of a point-like particle may be precisely known (as the state ψ above) but not localized in a single point.

The situation may be resumed in the following way. The states that seem, according to classical physics and our intuition, incompatible

---

[2]This is done in analogy with the term "Multiverse" accepted in the modern cosmology for the complicated geometry resembling many Universes existing besides with each other.

(alternative), may, according to quantum physics, coexist. In another formulation: *classical alternatives may be superposed, they may coexist.*

This strange, counter-intuitive feature of quantum systems (coexisting classically alternative states) has been experimentally proved for microscopic objects such as elementary particles, atoms etc. It is impossible to accomplish an experiment verifying whether this feature takes place also for macroscopic systems.[3] However, there is no reason why the same property might be invalid for macroscopic bodies as well. Moreover, it is impossible to make quantum mechanics logically closed if we reject universal character of this feature.

This feature (coexisting classical alternatives) is accepted to be universally valid in the variant of quantum mechanics suggested in 1957 by the american physicist Hugh Everett. According to the Everett's theory (Everett's interpretation of quantum mechanics), *classically incompatible states of our world may be in superposition, i.e. coexist.* In another formulation, coexist various classical worlds, which are called in this context "Everett's worlds". A cat may be alive in one Everett's world and dead in another Everett's world, and nevertheless these worlds coexist, they are parallel (this is a famous paradox of Schrödinger's cat, see Sect. 1.6.2.1).

The image of parallel worlds is more transparent than the image of superposed classically distinct states of the world. This is why Everett's interpretation of quantum mechanics is often called *"Many-Worlds interpretation"*.

The question arises naturally: if parallel worlds coexist, why then we see only a single world around us, but not a superposition of classically incompatible worlds. Why we see for example either alive cat or dead one, but not both of them coexisting in some sense or another? The answer given in the Everett's Many-Worlds interpretation is that the *alternative classical realities are separated by consciousness.* This means that an arbitrary observer perceiving the quantum world (all its "classical faces", or classical projections) perceives different projections independently from each other: in the picture of one classical reality there is no place for the others (although they are objectively not less real than the first one).

It is evident that, as a result of such a separation of alternatives by consciousness, we have an illusion that only a single world exists. Such is our subjective impression, even if objectively many parallel worlds coexist.

---

[3]Experiments with bodies consisting of 10^5 degrees of freedom confirm existing superpositions of classically incompatible states, but it is impossible to perform analogous experiments for usual macroscopic bodies which consist of 10^{23} degrees of freedom.

But in this case, is it possible to make sure that parallel worlds really exist? May be their coexisting is only "play of mind" of a theorist having no practical significance? We shall see that this is not true. Parallel worlds, of quantum reality, provides qualitatively new abilities to our consciousness. Actually it is only coexisting of parallel worlds that may explain mystical features of consciousness that have been noticed, investigated and exploited long time ago by various religions, oriental philosophies and psychological practices. Here we shall discuss this circle of phenomena in the framework of the natural sciences of European type. The special features of quantum mechanics make this possible.

For this aim we shall start from the assumption that parallel (Everett's) worlds coexist but are separated by consciousness. Then we shall make the next step that will finally lead to theory of consciousness including such features of it as super-intuition (direct vision of truth) and probabilistic miracles.

4.2   Consciousness: classical vision of quantum reality

4.2.1   *Consciousness as separation of classical alternatives*

Let us summarize what is our starting point. Quantum reality is presentation of the (objectively existing) quantum world with the set of classical worlds, or alternative classical realities (or simply alternative). Various classical alternatives are nothing else than "classical projections" of the quantum world. Nevertheless, subjectively an observer has an illusion that there is only one classical world around him. The reason of this illusion is that classical alternatives are separated in his consciousness so that they are perceived independently from each other. This is *classical vision of the objectively quantum world* (see Fig. 4.1).

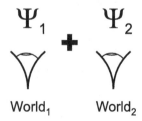

Fig. 4.1   Two classical realities (Everett's worlds) separated by consciousness

This idea, in one form or another, is already accepted by those who support the Everett's many-world interpretation of quantum mechanics [Squires (1994); Lockwood (1996); Whitaker (2000); Stapp (2001)]). We however shall make the next step and, unlike other authors, reinforce this proposition.

Let us assume [Mensky (2000a, 2005a)] that the relation between consciousness from one side and separation of alternatives from the other side is more than merely an association between two different phenomena or notions. Assume that these phenomena, which seem to be quite different (although related), are in fact identical to each other. In other words, we assume that *the separation of the alternatives should be identified with the consciousness*. We now specify this more precisely.

Everett's concept deals with two aspects of consciousness (see Section 3.6). The consciousness as a whole splits between alternatives, and a 'component' of consciousness lives within one classical alternative, perceives only this single alternative classical reality.

In psychology, only that which is subjectively perceived is termed the consciousness, i.e., only the 'classical component' of the consciousness, according to our terminology. Therefore, to identify the notion of 'consciousness' with some notion from the quantum theory, we must broadly interpret consciousness as something capable of embracing the entire quantum world (all alternative classical realities) rather than exclusively one its classical projection. Therefore, we arrive at the following *identification hypothesis*:

> The ability of a human referred to as consciousness is the same phenomenon, or notion, which appeared in (Many-Worlds version of) quantum theory as separation of the single quantum world into classical alternatives.

The identification hypothesis that we are now discussing is not entirely new. It is intimately related to those versions of Everett's interpretation which are sometimes given a separate name — the '*Many-Minds interpretation*' (see [Albert and Loewer (1988); Lockwood (1996); Vaidman (2002); Whitaker (2000); Zeh (2000)]). We believe that the proposed hypothesis is easier to apprehend and more fruitful.

4.2.2 *Consciousness is common for physics and psychology*

On the face of it, the step made when we adopt the identification hypothesis is not large. But it actually permits seeing the relation between the quantum measurement (observation of the quantum world) and the observer's consciousness in a radically different light. Wherein does the standpoint change when we identify the separation of alternatives with the consciousness? Previously, we knew that these phenomena, which belong to qualitatively different spheres, were nevertheless related to each other. We now believe that this is simply the same phenomenon.

An evident advantage of the resulting scheme is that it has simpler logical structure: instead of two poorly defined notions ("alternative separation" and "consciousness") we have only one. Even more important is that this single notion is characterized now from two qualitatively different points of view: in the context of psychology and in the framework of quantum theory. This allows to characterize this notion much better.

Then, two different spheres of knowledge (quantum mechanics and psychology) are now effectively unified with each other. Previously, the two spheres had no common elements (although there existed some functional relation between them), and now they have a common element, the consciousness. *The consciousness turns out to be the common part of quantum physics and psychology,* and therefore the common part of the sciences and the humanities.

Let us make this statement somewhat more precise. The common part of quantum physics and psychology, which may, in the context of quantum physics, be termed the separation of alternatives, is to be identified only with the deepest (or the most primitive) stratum of the consciousness. It is as if this consciousness stratum ly 'at the boundary of consciousness' and is intimately related to the effect of perception, i.e., to the transition from the state when one is not aware of something to the state when he has become aware of it.

To simplify the terminology, we just say that this common part of quantum physics and psychology is *consciousness.* Only sometimes, whenever necessary, we recall that we are dealing not with the entire diversity of phenomena commonly embraced by the term 'consciousness', but with the intangible that distinguishes the state in which a subject is aware of what is taking place from the state in which he is not.

The identification of consciousness with the separation of alternatives, i.e., of two phenomena from qualitatively different realms, explains why both of these phenomena are poorly comprehensible within the ordinary approach. The understanding is not achieved because each of these phenomena is analyzed only in the context of one realm and an important aspect lying in the other realm is omitted. Now, when we accept the identification, We have the benefit of possessing both senses of the complicated notion of consciousness.

In the subsequent discussion, we repeatedly rely on the notion of consciousness as the common part of physics and psychology, which allows us to present a clearer idea of the potentialities concealed in the philosophy of quantum mechanics. The resulting complex of ideas, based on the concept of Everett but differing from it by the identification of consciousness with the alternative separation, is called *Extended Everett's Concept* (EEC).

4.3 At the edge of consciousness

As it has already been argued, *Extended Everett's Concept* (EEC) is based on the identification of consciousness with the separation of (classical) alternatives (equivalently, separation of the parallel worlds). We shall see that this assumption leads to a number of very important conclusions about consciousness. The reason is very simple: if consciousness is identified with the separation of alternatives, then turning the consciousness off means disappearance of this separation, i.e. emergence of access to all alternative realities. The information from this enormouse "data base" makes feasible (in the state of unconscious) super-intuition, i.e. direct vision of truth.

Thus, extraordinary features of consciousness (and first of all super-intuition) should reveal "*at the edge of consciousness*" when the consciousness (i.e. the separation of the alternatives) disappear or almost disappear. What appears then instead of consciousness (in the usual understanding of this world) may be called extended consciousness, or *super-consciousness*.

Another very important assumption accepted in EEC is that consciousness has the ability to influence the alternative to be subjectively perceived. In a sense, this means that the ability exist to "control reality".[4]

Arbitrary as this second assumption may seem, it is nevertheless quite natural if the first one (identification of consciousness with the separation of alternatives) is accepted. Indeed, if human, with the aid of his consciousness, has access to the information of all parallel worlds (parallel classical realities), then, according to the general features of life, the means should exist for most efficient usage of this information for surviving and even for lifting of the qualtity of life of humans. Influence on the state of the environment, or arbitrary managing subjective reality, is the most efficient way to support the life level, and it must be provided.

---

[4]It is important to understand that only subjectively perceived reality is supposed to be controlled. The objective quantum reality (presented by the whole set of alternative classical worlds) is governed by the usual quantum-mechanical laws and cannot be arbitrarily controlled.

4.3.1   *EEC: Consciousness is the separation of alternatives*

The Everett's many-world interpretation of quantum mechanics and Extended Everett's Concept (EEC) make possible (and even almost necessary) the radical suggestion that, in some way or another, consciousness should have access to information contained in all alternative realities (all parallel worlds). This possibility is suggested by the fact that in Everett's concept, the consciousness as a whole (in contrast to its separate components) embraces the whole quantum world, i.e., all its 'classical projections'.

In the light of this circumstance, it is conceivable that the individual consciousness (or, in other formulation, a single 'component' of the consciousness), which lives in some Everett's world (in a definite classical reality), under certain circumstances may nevertheless, in some way or another, obtain information from the quantum world as a whole, i.e. to 'look into' other alternatives, other realities. In the Copenhagen interpretation this would be impossible, because no 'other' alternatives exist objectively. However, in the frameworkd of the Everett's Many-Worlds interpretation they exist, and the access to them is in principle feasible.

Let us consider this ability of consciousness first in the context of the Everett's many-world interpretation of quantum mechanics, and later on, make it more precese in the context of Extended Everett's Concept (EEC). It will be helpful to discuss previously classical character of the alternatives consisting the set of parallel worlds. Why are they classical?

4.3.1.1   *Why the alternatives are classical*

The classical character of alternatives is simply experimental fact. The world we observe around us is essentially classical. We always see some definite classical configuration of this world, and never observe anything similar to 'superposition', or coexisting, of classically distinct configurations (say superposition of a cat being alive and the same cat being dead). But why?

This question may seem incorrect. Nevertheless, it has a very simple answer in the context of the Everett's Many-Worlds version of quantum mechanics. If we recollect that the quantum world is *separated by consciousness* into classical alternatives, then it is almost evident that these alternatives should be classical. Separation just into classical alternatives is necessary because 1) consciousness is a property of living beings (say, humans), 2) classical state of the world is 'locally stable', i.e. the future of the restricted region of such a world depends only on its state inside this region or in a vicinity of it, and 3) local stability is necessary for local form of life.

Let us explain this a little bit. Why local life may exist only in a locally stable world? Because, by the very definition of life, living beings should be able to manage their surviving. This, in turn, means that they should have some (more or less efficient) strategy of surviving. But no such strategy can exist for local living beings in a locally instable world: they simply cannot predict what will happen tomorrow. It is only in locally stable (therefore classical) world that the future can be (with relatively good reliability) predicted and therefore strategy of surviving found.

Thus, consciousness separates the quantum world into its classical counterparts (alternatives) because (the only known for us) local form of life is feasible only in classical worlds. But there is only quantum, therefore locally instable world. Local life cannot exist in such a world taken as a whole. Yet local living being may live in separate classical components of this world. For this aim they should elaborate a special way of seeing this world. They should perceive separate classical components of the quantum world independently of each other, in order to live in these classical worlds independently of each other.

Such an ability to see the quantum world in its classical components should be a condition for any local form of life. For human beings this ability is called *consciousness*. For lower forms of (local) life the analogous ability may be called *pre-consciousness*.

In reality the set of many parallel living processes is performing in the set of parallel worlds, but 'subjectively' life in each of the parallel worlds is experienced. The reason is that each of these alternative worlds is locally stable.

4.3.1.2 *Accessibility to other realities*

This implies at least principal accessibility of 'other alternative realities' for consciousness (understood in a wide sense of the word).

Indeed, let us ask, whether one can, in some way or another, to 'look into' other alternative realities. Until we assume (as is commonly done in 'standard' interpretation of quantum mechanics) that all alternatives, except one, simply do not exist, then there is simply nowhere to 'look into'. But if all alternatives are equally real and the consciousness simply 'separates' their perception for itself, the fundamental possibility to look into any alternative, to become aware of it, does exist.

There is an image that illustrates the splitting of consciousness between alternative classical realities: the blinders put on a horse, such that it cannot look sideward and retains the direction of motion. In precisely the same way, the consciousness puts on the blinders, places 'partitions' between different classical realities in order that each 'component' of the consciousness would see only one of them and would make decisions in accordance with the information coming from only one classical (and hence relatively stable and predictable, i.e., livable) world.

However, just as a horse which is wearing blinders can nevertheless look aside by turning its head, so the individual consciousness, which lives in some definite classical reality, should most likely have, despite the partitions, the fundamental ability to look into other classical realities, other Everett worlds. Then, a man is able not only to imagine (which is certainly possible), but also to *directly perceive* some 'other reality', in which he might also live.

4.3.1.3　*The role of unconscious*

If we accept not only the Everett's many-world interpretation but also EEC, then it becomes even possible to qualitatively characterize the state of consciousness in which this can take place. It is possible to look into other alternatives (or, equally, to go out into the quantum world as a whole) only when the partitions between the alternatives vanish or become penetrable. According to the EEC, the emergence of partitions (the separation of alternatives) is nothing but perception, i.e., the emergence of consciousness, its 'origin'. And vice versa, the partitions vanish (or become penetrable) 'at the edge of consciousness', when the consciousness almost vanishes. Suchlike states are *sleeping, trance or meditation*.

Let us make a remark that is very important from the paractical point of view.

Starting with the hypothesis of identification consciousness with the separation of alternatives, we arrived at the conclusion that access to 'other classical realities', or other parallel worlds, appears "at the edge of consciousness" i.e., in the state when consciousness is (almost) turned off (sleeping, trance or meditation). The resulting way of getting information from the quantum world as a whole (i.e., from all parallel classical worlds) may be called *super-consciousness*. This however is only the simplest situation when the super-consciousness appears.

The ability called super-consciousness may appear even *when turned out is consciousness of some special aspect of reality.* Consider for example the following situation. A scientist is intensively working on a complicated problem and cannot solve it by all methods available for him. Making it all possible to clearly formulate the problem, he stops to think about it, thus turning his consciousness from this concrete problem. Then super-consciousness aiming at the given concrete problem starts to work. As a result, in some time a principally novel approach to the solution of this problem may unexpectedly come to the scientist as an instantaneous bright insight. This is nothing else as the super-intuition resulting from the work of the super-consciousness. It may arise even without turning off the consciousness completely, turning it off from the given problem is quite enough.

Moreover, it is shown by the adepts of oriental psychological practices that super-consciousness can be working all the time, even if the person is functioning in the usual regime. He may speak, listen other people and exchange replicas with them, make his usual everyday work, think and make solutions. However, all this time somewhere deep in his consciousness a point may exist in which consciousness is replaced by super-consciousness. This may be directed to a special idea or problem, it may expect some 'transcendent' sign or a piece of information from somewhere (in our context, from the quantum world as a whole).

In all such more complicated situations, the crucial role is played by the regime of unconscious. This regime may cover the whole area of consciousn or some directions of it.

The evident example is supporting of health. The regulations made for this are always performing in the regime of unconscious. Of course, most of these regulations are made with the help of "calculations" performed by the brain and other regions of the nervous system. However, there are good reasons to think that some regulations may be realized on the basis of the information obtained by super-consciousness. This must be in cases when the necessary information cannot be obtained by the usual ways.

For example, let some qualitatively new conditions appear in the environment, never experienced by the given organism and even the given species. Then there is no information in the brain or genetic apparatus about the optimal behavior in the given conditions. Then this information can be achieved with the help of the super-consciousness. For this various possible classical realities (particularly differing by the regulations of the given organism) must be compared with each other, and the most appropriate one (providing survival and even good quality of life) chosen.

Such very important "super-regulations" may perform in the regime of unconscious of one type or another. Some of them may be made when one is wakeful. The most important regulations of this type are surely performed in sleeping. This explains, first, why sleeping is so important in case of illness and, second, why long time without sleeping is killing human.

4.3.1.4 *Super-consciousness: Travel in time*

We mentioned the useful information obtained by super-consciousness in the quantum world as a whole (in the set of all parallel classical worlds). Let us make some point of this phenomenon clear. What is the nature of this information and why it is useful?

Consciousness (or rather subjectively experienced 'one component' of the consciousness) is living in a single classical world. Super-consciousness is going out in the set of all parallel classical world which realizes a single quantum world. The states of parallel classical worlds are superposed, and their superposition is the state of the quantum world. Thus, super-consciousness deals with the states of the quantum world. Therefore, we can say something about super-consciousness if we reall the laws of quantum world.

The main feature of states of any quantum system (and therefore of the quantm world) is following. The evolution of any quantum state is governed by the Schrödinger equation, and the state in some time moment unambiguously determines the state in all other time moments in the past and in the future. The state of the quantum world is therefore defined in all times.

This is why super-consciousness, being in the quantum world (accessing all parallel worlds), has information not only about the present time of this world, but about it states in all times. Therefore, it may "see" non only "now" of all parallel classical worlds, but their past and future. From this information should be clear what of these worlds (what of the alternative classical realities) is advantageous. Say, comparing two of the alternative realities and following in their future, it is seen that in the first reality the person in future is dying (or heavily ill), and in the second reality he is alive and health. It is evident that the first reality is advantageous.

In more convenient wording, the super-consciousness may compare with each other various *"Everett's scenarios"* (the chain of the classical realities in the sequence of time moments). It can follow each of the scenarios up to far future and see in what of them the state of the given person is better.

This allows the super-consciousness to judge about what of the Everett's scenarios is advantageous.

4.3.2 Subjective probabilities and probabilistic miracles

In the framwork of Extended Everett's Concept we accept, besides the identification hypothesis, one more important assumption. We shall assume that consciousness has the capability to influence subjective probabilities of the classical alternatives. This means that consciousness can menage subjective reality i.e., the subjectively perceived reality. What does this mean?

Let I am going to the countryside tomorrow and am interested for the tomorrow whether to be sunny. According to the Everett's Many-Worlds interpretation the whether tomorrow will be sunny in some of the parallel worlds and runny in the others. Objectively my consciousness will be splitted between these worlds, or, in the other words, in each of the parallel worlds a replica of myself will be present.

However it is reasonable for me to ask: what of these two realities I shall perceive subjectively. The answer may be only in terms of probabilities, and the probability for me to perceive a definite reality may be called the *subjective probability* of this reality.

The usual answer to my question about the subjectively perceived reality is following: calculate the probability of each of these variants of the whether (or listen radio for the meteorological forecast). This will give you the *objective probability* of each of the two alternative realities (sunny or ranny day). Then the same estimate will be valid also for the subjective probability. In this usual argument the subjective probability of a given alternative reality is equal to its objective probability. But is it actually necessary?

We shall assume that the subjective probability may differ from the objective one. Moreover, we shall assume that one can influence the subjective probability to make more probable that the subjectively perceived reality be advantageous for him (in our example, I can make the sunny whether more probable). Strange as this assumption may seem, it is quite natural after the consciousness is supposed to be identical to the separation of the alternatives.

4.3.2.1 *Objective and subjective probabilities*

According to Everett's interpretation, there exists an infinite set of Everett's worlds ('classical realities'), each of which is characterized by some probability (or, in the case of a continuous set, probability density). The probabilities may be calculated according to conventional quantum-mechanical rules.[5]

Probabilities of alternatives found according to the quantum-mechanical rules are commonly considered as objective, so that no other probabilities are considered. The objective character of these probabilities is experimentally verified in the following way.

An experiment with a simple physical (microscopic, therefore quantum) system is performed repeatedly. Such an experiment may be considered as the evolution of the system with the given initial state and the measurement of some characteristic of this system at the end of the experiment. According to quantum mechanics, the measurement at the end of such an experiment may give various results (records) that correspond to *various classical realities* of the Everett's Many-Worlds interpretation.

Each of the measurement results can be characterized by the definite probability found according to the quantum-mechanical rules. The objective character of the probabilities is confirmed by the fact that each measurement result is obtained with frequency which is proportional to the corresponding probability (more precisely, the frequencies become close to the probabilities in the limit of very large number of the experiments).

We can also introduce *subjective probabilities* of the various measurement results in a single experiment or measurement. The subjective probability is the measure of whether it is reasonable to expect that the given result will be obtained in the given experiment. In the framework of the Everett's Many-Worlds interpretation, one can introduce the subjective probability of various classical realities. Equivalently, subjective probability of one of the parallel worlds may be defined as the probability for the given person to find himself in this world.

In principle, objective and subjective probabilities of some event may differ. In case of series of simple physical experiments, it seems intuitively reasonable to consider the objective (calculated) and subjective (expected) probabilities coinciding as the consequence of the experiment. However,

---

[5]In the simple example (considered in Sect. 3.2) with two alternatives $|\psi_1\rangle$ and $|\psi_2\rangle$, these are the probabilities $p_1 = |c_1|^2$ and $p_2 = |c_2|^2$.

even in this simple case it is not evident (we shall discuss this later in connection with "probabilistic miracles").

Even less ground exist to identify these two types of probabilities in the case of complicated events that cannot be repeated. Take for example such events as the disease or health of the given person, the crash of the given car at the given time moment, or the explosion of a supernova in the given region. Indeed, we cannot organize the series of experiments with the given person in the given conditions because the conditions are changing continuously. We cannot replace such a series by the statistics obtained in the observations of various persons because the human beings are not elementary particles and are not therefore identical.

Thus, we shall accept the (more general and therefore more reliable) assumption that the *subjective probabilities* of observing various events, or perceiving various classical realities (being in various parallel worlds) *are not necessarily equal* to the corresponding *objective probabilities* calculated with the help of the natural scientific laws.

Our aim is to justify the assumption that the subjective probabilities of various classical realities ay be influenced by consciousness. We shall do this in the context of Everett's Many-Worlds interpretation of quantum mechanics. This assumption (fantastic as it seems) is not quite new. Several authors have hypothesized (not necessarily in the framework of Everett's interpretation) that the consciousness can affect the probabilities of various alternatives [Squires (1994); Eccles (1994); Beck and Eccles (2003)].

4.3.2.2 *Subjective probabilities in terms of parallel worlds*

In the context of the Everett's Many-Worlds interpretation, the probability of a given alternative is quite often interpreted as the fraction of those Everett's worlds in which this alternative is realized. In this case it seems natural and even necessay to interpret this fraction both with objective (calculated) and subjective (expected) probability which therefore should be equal to each other. But is this actually necessary?

No, in principle, there remains the possibility to consider the objective and subjective probability to be not equal. The reason is that the "fractions" of various types of Everett's worlds is not always well defined numbers. The "fractios" may be defined in many different ways. One of these ways may give the objective probability distribution, many others may correspond to the subjective probabilities associated with various persons (observers).

Let us consider this issue of "fractions" of various types of the Everett's worlds in conection with the concept of subjective probabilities.

The conclusion that the probability distribution of alternatives is unambiguously defined by quantum-mechanical laws would be beyond doubt if the selection of one of the alternatives were among those physical laws that are objective and independent of the observer's consciousness. But in the framework of Everett's concept *the separation of alternatives is performed by the consciousness* (or, in EEC, even more definitely: the separation of alternatives is the consciousness). It would appear reasonable to suggest that the consciousness can affect not only the character of alternatives but also their *subjective probabilities*, i.e., the probabilities of which alternative will be subjectively perceived.

It is thus reasonable to assume that the consciousness can increase the probability of finding its way into those classes of Everett's worlds that are preferable to it for some reason.

This assumption may seem to be unacceptable when the probability of an alternative is identified with the fraction of Everett's worlds of the corresponding type (in which this alternative is observed). On the face of it, the number that expresses 'the fraction of the worlds of a given class' should be universal, and must then coincide with the quantum-mechanical probability. Then it may not be different for the consciousness of one observer or another (may not be subjective). Were the number of Everett's worlds finite, this would indeed be the case.

However, the very notion 'the fraction of the worlds of a given class' is meaningless for the infinite set of worlds, and the argument added in favor of the universal character of the probability distribution loses its force.[6] That is why in the case of an infinite set of Everett's worlds, defining different probability distributions on this set is quite admissible and the assumption of *the effect of consciousness on the (subjective) probability distribution is not self-contradictory*.

To make this statement pictorial, we assume that an infinite set of the observer's 'replicas' one after another are sent by the consciousness into Everett's world of one type or another in order to fill the infinite set of worlds. The situation is demonstrated by Fig. 4.2.

For simplicity, we assume that there are only two alternatives, i.e., two world types. Then, for one observer, the replicas may make their way into the parallel worlds in such a way: one replica into the type-one world and

---

[6]The deep mathematical reason of this is that an infinite set possesses a paradoxical property: it may be put in a one-to-one correspondence with its own subset.

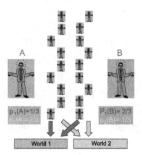

Fig. 4.2 Two observers have different preferences and influences the subjective probabilities of the two parallel worlds directing more if their "replicas" into the worlds they prefer.

the next two into the type-two world, then again: one replica into the type-one world and the next two into the type-two world, etc. This corresponds to the following: the probability that a given replica find its way into the type-one world is equal to 1/3 and the probability to find its way into the type-two world is equal to 2/3. The consciousness of the second observer may send its replicas into the same worlds differently: initially, two replicas into the type-one world and the next one into the type-two world, then again two replicas into the type-one world and one into the type-two world, etc. As a result, for each of them, the probability of finding its way into the type-one world is 2/3 and the probability of finding its way into the type-two world is 1/3. However, both the above procedures have the effect that *each Everett's world harbors one replica of each of our observers.* Clearly, it then makes no sense to pose the question what is the fraction of type-one worlds (because there are infinitely many worlds).

This reasoning does not prove, of course, that the consciousness can indeed control probabilities but it shows that this assumption is not self-contradictory, and we shall include this assumption into EEC. We thus assume that *the consciousness can make some event probable even though the probability of this event is low according to the laws of physics* (quantum mechanics). We have to make an important improvement on this formulation underlining that *the probabilities are subjective:* the consciousness of a given observer can make it probable that he will (subjectively) see this event. This subtle correction is necessary because, according to the Everett's Many-Worlds interpretation (and therefore in the context of EEC) all parallel worlds objectively exist.

4.3.2.3   *Probabilistic miracles*

When an event whose probability is extremely low according to the laws of physics is made probable by the consciousness, the event taking place may look like a miracle. Yet, this is the miracle of a special type that can be called *probabilistic miracle*. The probabilistic miracles are consistent with the physical laws, *they do not contradict science!* The deep reason of this is that the result of any measurement is, according to quantum mechanics, a random event. The prediction of what will be observed may be only probabilistic.

By the way, one of the starting point for the collaboration of Pauli and Jung on the topic of connection between quantum mechanics and consciousness was the phenomenon of strange inexplicable coincidences observed by Jung in his work as psychologist. He called these coincidences *"sinchronisms"* because the strangeness of the observed coincidences was not in the events that happened but in that they happened simultaneously. In the framework of our approach, sinchronisms may be interpreted as a sort of the "probabilistic miracles".

It is significant that there exists one absolute limitation for probabilistic miracles. If the probability of some (mentally constructed) 'classical reality' is equal to zero (i.e., this reality is actually absent among all possible alternative classical states of the world), the individual consciousness cannot make the (subjective) probability of finding its way into this reality nonzero. The subjective probability is in this case equal to zero. The corresponding alternative cannot be observed.

Hence, not every miracle is possible. That which is absolutely forbidden by physical laws (that which takes place only in fairy tales) cannot be realized in any case. And that which is unlikely but possible can be realized 'in reality', even though the probability calculated by physical techniques is very low.

4.3.3   *More precise formulations and examples*

We make two brief remarks, which are required for the correct understanding of the heart of the problem.

The first remark is intended to specify the interpretation of the hypothesis about 'identification' of the consciousness (commonly considered in the framework of psychology) with the separation of alternatives (the notion of quantum physics). According to this hypothesis, the consciousness (= the separation of alternatives) is a common part of psychology and quantum physics. It becomes possible to study two aspects of this object, the consciousness, to view it from the standpoints of two areas of knowledge different in character: from the standpoints of physics and psychology.

Of course, in doing so we see this object differently, and different features of this object turn out to be significant. In physics, to characterize our subject (the separation of alternatives), we consider the simplest experiments with the simplest objects (say, elementary particles or atoms), which were intentionally selected among the primitive ones such that they are amenable to investigation by mathematically accurate methods. When dealing with the same subject (now called consciousness) as viewed from the standpoint of psychology, we face substantially more complicated and far less clearly defined complexes (such as perception of the world by human beings or states of their bodies in the complicated environments).

This is of significance, for instance, when we are dealing with the hypothetical possibility of affecting the selection of an alternative with the aid of the consciousness. It is unlikely that the consciousness can have an appreciable effect on what the localization of an electron turns out to be or in what direction it flies. If the consciousness does have the ability to affect the selection of reality, this most likely applies to those aspects of this reality that are vital to the person (because, according to our reasoning, this very phenomenon, the consciousness, emerges due to its being vital to living creatures).

If, for instance, a close relative dies in one of these realities and remains alive in another, the conscious subject is highly motivated to select the latter alternative. If he believes in this case that he is able to affect the selection of reality, it is not inconceivable that he will actually increase the probability to some extent that he will witness precisely the latter alternative (whether suchlike possibilities should be used is a separate question, and the answer is not as obvious as it might seem to be).

The 'identification' of the separation of alternatives in quantum physics with the effect of realization (becoming aware) in psychology should be understood with this reservation only. Only the deepest layers of the corresponding phenomena are identified, their underlying principle but not their manifestations, which may be extremely unlike in the realms of physics and psychology.

The second remark concerns the new prospects in psychology and in the humanities in general that stem from its relation to physics. We say that the consciousness (the psyche) may, in the framework of the concept involved, have certain features that are not ascribed to it in 'classical' psychology (such as the ability to leave the classical alternative and enter the quantum world, i.e., look into other, alternative, realities or even affect the selection of 'its own' reality). Of course, these hypothetic possibilities call for verification.

However, it would be quite reasonable to attempt to identify these 'new' possibilities with those extraordinary phenomena in the field of psychology, the theory of consciousness, and the psychological practice, which have long been noted, studied by different methods, and even exploited. From this standpoint, the 'new' features of consciousness under discussion might have been known for a long time. If so, some facts in support of the concept under consideration may be found without any additional verification. But in this case, too, careful and cautious work is required to analyze the known facts and to compare them with what might be expected in the framework of Everett's concept.

Among the extraordinary phenomena in the realm of consciousness (psyche) that may be related to our concept, we mention, first, special (trancelike) states of the consciousness, the state of the consciousness during sleep in particular, and, second, nonverbal and uncontrollable thinking, which plays an important part in science and which is explicable, in the view of Penrose [Penrose (1991)], Ch. 10, on the basis of quantum physics.

Very much has been said and written about the special states of consciousness and the state of sleep (in this connection, see the intriguing "Ikonostas" essay by Pavel Florenskii [Florensky (1996)], pp. 73–198). The phenomenon of nonverbal thinking is less known. We briefly explain what is implied.

The thinking of a scientist is commonly believed to be a strictly logical and consistent flow of ideas, which are committed to paper or, at any rate, can be stated on paper when wanted with the aid of our ordinary language (with the addition of a number of formulas and drawings). And this is indeed the case at the initial stage of work, when the problem is formulated, and at the final one, when the result is formulated. But the key stage of the scientist's work, which actually yields the result, is the insight. And it turns out that the scientist's thinking at this stage quite often (and maybe always) assumes a nonverbal form and proceeds in an uncontrollable manner, independently of his will (however, after the intensive and completely controllable work at the preceding stages).

Roger Penrose, in his book "The Emperor's New Mind" [Penrose (1991)], provided examples of important discoveries made in a nonverbal form. Maybe the most striking fact about the testimonies of the great scientists he cited was that at the instant of discovery, in the absence of formal proofs of the verity of their insight, they were absolutely certain that it was true.

This extraordinary and at the same time extremely important phenomenon is impossible to explain in the normal way. It seems to be attributable to the fact that the consciousness enters the quantum world at that moment. Of course, much remains to be done in this area, but some preliminary considerations suggest themselves immediately. In particular, the idea that a scientific discovery is made 'at the edge of consciousness' leads to the following practical recommendation.

After a period of intensive preliminary work on a problem, at the instant when it is required 'to guess the key to its solution, it is expedient to 'switch the consciousness from this problem for a time to something else (either to another problem, or even simply to relaxation). In this case, the work on the problem actually goes on, but on the level of subconsciousness (or, makin use of more adequate terms, in the regive of *super-cognition*, or *super-consciousness*), which is required for the 'discovery', i.e., for the emergence of qualitatively new considerations on the problem. The high efficiency of this procedure has been proven in practice. Similar recommendations are quite frequently encountered in the literature on scientific methodology.

And this is just one example most kindred to a representative of science. There are many other amazing phenomena in the realm of consciousness, and many of them are supposedly authentic.

4.3.4 *Relation to religion and oriental philosophies*

Thinking of extraordinary phenomena that are in one way or another related to human consciousness, we have to mention those forms of their cognition, or even controlling them, that are not scientific. First and foremost, these are different religious beliefs and oriental philosophies. Scientists are fully tempted to exclude this area of human thought as being unscientific, i.e., unreliable. However, one can hardly wave away the doctrines that has existed for millennia and represent may be the most stable phenomenon in the sphere of spiritual human life. This stability is most probably an indication that all these unscientific notions rely on something actual, even though their actual basis is frequently put in a fantastic form to strengthen its emotional action.

Of interest from this standpoint are oriental philosophies, which directly encourage their adepts to work on their own consciousness. We believe that *Buddhism, Daocism* and similar beliefs are most interesting in this respect (see Ref. [Mansfield (1991)] about the substantial conceptual proximity of quantum mechanics to 'Middle Way Buddhism').

There are at least two important features of this philosophical–psychological school that seem attractive from the standpoint involved.

First, Buddhism does not require blind belief in the dogmas it proclaims. Disciples are urged to believe only when they assure themselves in the course of work on their own consciousness that the doctrine is correct. Second, Buddhists consider their task to learn to perceive a special state or sensation, which is impossible to exactly express by words and which may be characterized approximately as 'the root of consciousness', 'the origin of consciousness', or 'the preconsciousness'. This is an elusive state that precedes the emergence of consciousness. Learners are urged to work on their consciousness until they catch this sensation of 'being between the consciousness and the absence of consciousness'.[7]

It is easily seen that the state of consciousness which is the goal of Buddhists bears much resemblance to the deepest or most primitive layer of the consciousness (being "at the edge of consciousness"), which is identified with the separation of alternatives in our Extended Everett's Concept.

4.4   The need for the new methodology

Starting from the Everett's concept and identifying, in the framework of Extended Everett's Concept (EEC), the consciousness with the separation of alternatives, we see that the consciousness may have extraordinary properties: the capability for looking into 'other classical realities' and even affecting the selection of the reality in which it lives. It is significant that these features of the consciousness, if they do exist, are in principle observable, they can be discovered and investigated. The Extended Everett Concept may therefore be verified, i.e., confirmed or refuted, by way of observations. Thus, the most significant drawback to Everett's original interpretation, namely the impossibility to verify it experimentally, is in principle overcome in EEC.

---

[7]Let us remark in this connection that the meditation technique, which is rather well known to Europeans, is commonly treated as the skill of switching off one's consciousness, but its true sense is to learn to be between the consciousness and the absence of consciousness.

However, verification of the EEC requires change in the methodology accepted in physics. The purely objective approach to the observed phenomena is inappropriate, and the verification should take into account, in some way or another, subjective aspects of what is observed.

4.4.1 *Inclusion of subjective*

It should be realized that the verification of EEC would be quite unusual and would not fall into the pattern of conventional physical methodology. The point is that *the verification implies the observation of an individual consciousness.* Let us assume that these observations turn out to be in agreement with the predictions of the Extended Everett's Concept. Would this be proof of the verity of this concept from the standpoint of physics and physicists? It is no means evident. From the other side, subjectively it may be very convincing, and this may be the only thing that makes sense.

In physics (and in natural sciences in general), it is agreed that only series of experiments with repetitive results are truth criteria. Moreover, these experiments are to be carried out by different experimenters (to confirm the objectiveness of the experimenters and independence of the experiments results from the details of the experiments setup related with the person conducting the experiment). Experiments on one's own individual consciousness or observations of this consciousness lack probative force from this standpoint.

To illustrate the originality of the situation, we consider in greater detail what should be expected if the assumption is true that consciousness can affect the subject probabilities of the various alternatives (i.e. that the probability to observe the given alternative may be arbitrarily increased or decreased).

If this assumption is actually valid (as we suggest), the consciousness can make significant the *subjective probability* of some event even though its *objective probability* (one calculated by the usual scientific methods) is small. In case of not simply small but negligible objective probability (say, 10^{-10}) this may look like a miracle (we call this class of events *probabilistic miracles*).

As noted in Section 4.3, zero objective probability may not convert into a nonzero subjective probability. This means that the consciousness can make probable only those events which may take place without influence of consciousness, in the natural way, according to the usual physical laws. This turns out essential for the analysis of the situation.

This is almost evident. Indeed, let us assume that some person actually does have the capacity to ensure, by the effort of his will (his consciousness), the course of events he likes. Then he would never be able to guarantee absolutely clearly that it was really he who has so affected the course of events. Even if he has many times ensured the realization of unlikely events ('work a miracle'), *there always remains the probability that the events have taken this course in a 'natural way'*, in accordance with the ordinary laws.

Therefore, even if 'probabilistic miracles' are possible, the evidence that these are indeed 'human-made miracles' and not good luck will never be absolute. And therefore, anyone who decides not to believe in them would have grounds to do so. A skeptic would have the opportunity to doubt even on finding himself, together with the miracle-worker, in that Everett's world (in that classical reality) where the unlikely event was realized.

Moreover, the 'unbeliever' himself would prefer to find himself in the world where the 'miracle' does not occur. If he also influences the subjective probabilities, he would prefer to see the realization of that alternative where the 'miracle' does not happen. For the skeptic, therefore, the probability that he sees the an unlikely event ('miracle') with his own eyes remains low.[8]

Thus, if it is assumed that consciousness can modify the alternative probabilities, the situation appears to be very strange. Those who believe in this assumption will have, with an appreciable probability, an opportunity to make certain that it is correct, i.e., that the consciousness does affect the probabilities of events. Those who are unwilling to believe this, with a high probability will make certain that this does not take place. Skeptics will find themselves in those Everett's worlds where ordinary physical laws, objective and consciousness-independent, are valid. But those who prefer to believe in consciousness-worked 'miracles' will find themselves in those of parallel worlds where such 'probabilistic miracles' do occur.

When considering the assumption of the effect of consciousness on the subjective probabilities of alternatives, one is forced to accept the fact that *the problem of truth criteria should be considered with much greater caution* than is generally accepted in natural sciences. This has the following implication: either the Extended Everett's Concept cannot be included into the realm of physics (and of natural sciences in general), or *the methodology of these sciences should be substantially broadened.*

---

[8]Here, we are dealing with one of the issues that is counterintuitive and therefore difficult to understand. That is why great care is needed in the analysis of a situation where the results of the effort of a 'miracle-worker' are observed by other people, among which are those who are inclined to believe him and skeptics who are unwilling to believe.

The new methodology should, first, allow, as the instrument of theory verification, the experiments involving individual consciousness or observations of it. This methodology should, second, consider the possible effect of a priori aims (inclinations) on the results of observations.

This would be actually a novel methodology. Let us remark, however, that a detailed analysis shows that even without the "radical" assumption of the role of consciousness, in the framework of conventional scientific methodology, the inference about the truth always relies on a series of intuitive judgments whose role is commonly not realized in full measure [Feinberg (2004)].

It would be very strange if the Extended Everett's Concept with those new entirely unexpected possibilities it promises had to be rejected only because it proves to be incompatible with the presently existing scientific methodology. Work in this area will most likely be continued if the above-noted possibilities are borne out.

The situation that may arise in this case is perhaps similar to the situation that formed when non-Euclidean geometries were proposed. These new geometries were incompatible with the methodology accepted in mathematics at that time: they necessitated the abandonment of the fifth Euclidean postulate, which was treated as indispensable in geometry. However, it was extremely interesting to pursue the new direction, which opened up quite unexpectedly, and see what came out of changing the methodology by abandoning the fifth postulate. And that opportunity was not missed, of course. It is most surprising that before long, the speculative geometries constructed on this path were endowed with real embodiments, and then this gave birth to the amazingly beautiful and splendid geometrical world that bears the name of General Relativity and proved to adequately describe our Universe.

4.4.2 *Only subjective is important*

The main concern of any professional physicist is in convincing other people (and often himself too) that the results of his investigation are valid, i.e., in agreement with the *objective reality*. However, the above analysis clearly shows that the purely objective reality does not exist. At least in observations (measurement results) *subjective aspect cannot be distinctly and unambiguously separated from the objective aspect*.

This shows up in the above complicated situation with the verification of whether managing reality is possible or not. No objective judgment is possible as to whether probabilistic miracles can happen or not. Any judgment about this should include subjective element. It is impossible to purely objectively prove or disprove existence of probabilistic miracles.

Just this indeterminacy enables unification of the area called (somewhat arbitrarily) consciousness to belong to both the sphere of natural sciences (quantum mechanics) and spiritual knowledge (psychology and psychological mystics). This is illustrated in Fig. 1.2 on page 14.

This impossibility to prove something (say, probabilistic miracles or super-intuition) objectively may bitterly disappoint some physicists. However, this is not so important for people in their everyday life. If something helps them in their life and being helpful is confirmed by repeated experiences, people do not very much think about the ontological status of these things. They only make use of them.

This is why most of physicists do not accept the sphere of mystics and consider mystical phenomena to be principally impossible, but people far from natural sciences easily accept this sphere if their practical experience evidences, in some way or another, that it exists and participates in the important events of life. This is why religion is the belief of so many people despite of the authority of science.

One may say that subjective is the only important in life of 'simple' people. The approach presented in this book and based upon Extended Everett's Concept, shows that the confrontation between science and spiritual teachings have no real grounds. Moreover, the central part of the natural sciences, quantm mechanics, cannot be made logically closed and cannot be actually understood without inclusions the spiritual notions into it.

4.5   Quantum correlations and telepathy in EEC

We considered above the two phenomena of mystical character observed revealing themselves in consciousness, and their explanation in the framework of Extended Everett's Concept (EEC). These phenomena are super-intuition and (probabilistic) miracles. There is one more mysterious phenomenon that is often experienced. It is *telepathy*, i.e., mental influence of one person onto another (in particular "reading thoughts"). The typical example is panic of a mother at the moment when her son is in the lethal risk thousands miles from her.

There exist explanations attributing telepathy to the quantum correlation of living organisms [Josephson and Pallikari-Viras (1991); Villars

(1983); Lovelock (1990)]. According to this hypothesis, telepathy arises as the manifestation of quantum non-locality (see Sect.3.4.1). The difficulty in this case is following. The effect of quantum non-locality arises only for the systems evolving in the quantum-coherent regime when no decoherence takes place. But decoherence is caused by any non-controlled interaction of the system with its environment. It is unclear how quantum-coherent regime may exist in the human body, say, for some material structure in brain.

It is natural to ask what may be additionally said about this hypothetical effect in case if we accept EEC. Does the specific features of EEC make more plausible appearance of quantum non-local effects leading to telepathy. It turns out that EEC naturally lead to the conclusion that the effects of non-locality must exist, but for the phenomenon of super-consciousness instead of the states of restricted material systems.

Indeed, in EEC we consider the whole world as a quantum system. This is principal. The system under consideration is actually the whole world but not its restricted part. If considering a part of the world we have something lying outside this part, in its environment. In this case the system under consideration may be treated as being observed by its environment. As a result, this system decoheres, so that its quantum features partly disappear (see Sect. 3.3). One may say that some of quantum correlations disappear in this case converting into the classical correlations.

In particular, no material system in brain can be correctly considered as evolving in purely quantum-coherent regime. The effects of quantum non-locality (and telepathy) between the material systems in two different brains are impossible in this case. The reason is that each such system is restricted and undergoes decoherence (and effectively becomes classical) under the influence of its environment.

Contrary to this, we consider, in the framework of EEC, not restricted material systems in brains but the whole world. There is no environment for this special system, and the world as a whole does not decohere. It evolves in the purely quantum-coherent regime, and its quantum correlations never convert into classical ones.

Let us formulate this more precisely. Telepathy arises as an effect of quantum non-locality. The necessary condition for this is the purely quantum regime, i.e., quantum coherence, absence of decoherence. This condition is realized not in consciousness, but in super-consciousness which is nothing else as the state of the quantum world as a whole, it being (in all times and in all space regions). Super-consciousness is simply a tem for the way of being of the quantum world.

This is why consciousness of different people may differ, but the super-consciousness is common for all of them. The consciousness of each person touch the (common) super-consciousness "at the edge of consciousness". And if the consciousness of a given person touch the super-consciousness, it have access to the consciousnesses of all other humans. It is clear that the explicit "perceiving" of the consciousness of another person (telepathy) may arise only in the special case when these two persons are connected with each other in a special way (e.g. by strong emotions or being relatives). Then the consciousness of one of them is targeted by the consciousness of the other.

This issue may be shortly formulated in the following way that underlines the main points of it. Coherent superposition of the states of a macroscopic system (to say nothinh of the whole world) cannot be distinguished from the mixture of the same states with the help of any real devices. However, the difference objectively exists, and *such a "super-device" as consciousness* can distinguish these two situations. The deep reason is that these "super-devices" include the whole world, not a restricted part of it.

Consideration of infinite systems (as the whole world in the present case) is a principal point of the approach exploited in EEC. We saw this already in Sect. 4.3.2.2 when considering subjective probabilities. It has been shown there that various probability distributions (objective and subjective) may exist only because of infinite number of Everett's worlds. In case of finite number of parallel worlds the subjective probabilities had to be equal to the objective ones.

4.6   Conclusion

The Extended Everett's Concept (EEC) formulated here allows to explain the nature and mystical features of consciousness. This approach originates from the conceptual problems of quantum mechanics and provides a higher level of understanding (or philosophical elaboration) of this branch of science. The situation is unique. It seems that the last tendencies in quantum mechanics including EEC realize the final stage of the great scientific revolution starting by the creation of quantum mechanics but not yet finished because the conceptual problems of quantum mechanics are not finally resolved. Their final resolution includes unification of natural sciences with the sphere of spirit.

In Chapter 7 we shall develop an alternative approach tO the problem that has even wider circle of applications giving, instead of the explanation of consciousness, the Quantum Concept of Life (QCL).

4.6.1 *The problem of the century*

The situation with conceptual problems (paradoxes) of quantum mechanics is unique. It will soon be a century since the problem of quantum paradoxes (the 'measurement problem' as it is often called because the paradoxes arise in connection with measuring quantum systems) has remained unsolved; however, time and again, on an increasingly broader basis, it is confirmed that the problem still exists and remains to be solved (see, e.g., Refs [Markov (1991)]; Ch. 1 in [Ginzburg (2003)]).

This situation most likely signifies that the solution to the problem is to be found in a quite unexpected direction or that the character of the solution will be unusual from the standpoint of stereotypes formed in physics. That is why, when estimating the solutions being proposed, one should always be prepared to encounter unexpected solutions. This would prevent rejecting emerging shoots of truth for the reason that they seem unusual.

That the problem is nontrivial can be confirmed by viewing the list of great scientists engaged in the problem (we refer to Bohr, Einstein, and Schrödinger [Bohr (1949); Schrödinger (1944)], to say nothing of Heisenberg, Pauli, and Wheeler, to name but a few).

4.6.2 *Solution through the Everett's concept*

In our view, the solution may be attained in the direction outlined by Everett's "Many-Worlds" interpretation of quantum mechanics, which has attracted considerable attention in the last decades (see review Ref. [Vaidman (2002)] and the references therein). It is not only abstract problems such as 'the measurement problem' that are involved. In the context of new quantum-mechanical problems, in particular the theory and practice of quantum computers, some researchers (for instance, David Deutsch [Deutsch (1997)]) resort to Everett's concept as a convenient language for specific investigations. Of course, this is highly subjective, and the majority of physicists employ conventional quantum-mechanical language even in the area of quantum informatics. However, in conceptual problems, Many-Worlds interpretation supposedly furnishes a new quality.

We believe that a helpful viewpoint — in attempts to solving 'the measurement problem' on the basis of Everett's interpretation — is in our Extended Everett's Concept (EEC).

4.6.3   Main points of EEC

The approach called Extended Everett's Concept (EEC) emerges when the consciousness is not merely functionally related to the separation of the quantum world into Everett's alternative worlds (as in the original Everett's interpretation of quantum mechanics),[9] but is completely *identified* with it. In EEC a logically consistent chain of reasoning reliant on this identification is constructed (see Fig. 4.3; details may be found in Section 4.2 and Ref. [Mensky (2000a)]).

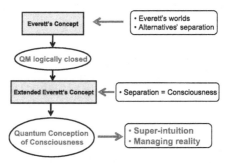

Fig. 4.3   Logical chain from quantum mechanics to consciousness

It is significant that thus extended concept may in principle be verified by *observations of an individual consciousness*. Close (coincident in some points) constructions have been undertaken by several authors, as is seen from the references cited. Especially much has been written (both in the framework of Everett's interpretation and beyond it) about the relation between the consciousness and the state reduction (remark although that no reduction is assumed in EEC).

To summarize the foregoing, the main points of the Extended Everett Concept and the naturally ensuing consequences can be formulated as follows:

---

[9]In a more precise wording, the alternative Everett's worlds are separated by consciousness not in the original Everett's 'Many-Worlds interpretation but in its variant often called "many-minds" interpretation (see [Albert and Loewer (1988); Lockwood (1996); Vaidman (2002); Whitaker (2000); Zeh (2000)]).

(1) The set of alternatives which is characteristic of the quantum notion of reality (resulting from the analysis of quantum theory of measurements) is interpreted as the set of projections of the quantum world referred to as Everett's worlds.

(2) Separation of the quantum world into alternatives is identified with the human function termed the consciousness.

(3) The classical nature of every alternative into which the quantum world is separated by the consciousness is determined by the fact that it ensures the stability and local predictability of the ambient world, as perceived by the consciousness, which is the necessary condition for life.

(4) In special states (on the verge of unconsciousness), consciousness gains access to the quantum world as a whole, beyond a single classical projection. This ability called super-consciousness may account for the extraordinary phenomena of super-intuition and probabilistic miracles observed in the realm of the psyche. These phenomena play the central part in the nonscientific forms of cognition of spiritual human life (oriental philosophies, religion, mystical doctrines).

Item 3 in this list is most important. It explains why in the measurement (perception), there occurs splitting of the quantum world into precisely the classical alternatives. The splitting of the quantum world into 'classical realities' (which are actually the mere projections of the solely real quantum world) turns out to be the necessary common property of all living creatures, i.e., the definition of (local) life.

In this connection, we note that in his article "What problems of physics and astrophysics seem now to be especially important and interesting at the beginning of the 21st century?" [Ginzburg (2003)], Ch. 1, V L Ginzburg names, among the three 'great' problems, both the interpretation problem of quantum mechanics and the problem of reductionism, i.e., the question of whether the phenomenon of life can be explained on the basis of presently known physics. We have seen that Everett's concept naturally combines both these problems and in a sense reduces one to the other (although there is no reduction in the direct meaning of this word).

Moreover, the last of the three 'great' problems mentioned in Ref. [Ginzburg (2003)], Ch. 1, namely the issue of entropy increase, irreversibility, and the 'time arrow',[10] may also bear relation to Extended Everett's Concept (see Chapter 6 for details).

---

[10] In V L Ginzburg's list, this problem is enumerated first.

The point is that, according to EEC, the quantum world as a whole (not splitted in classical alternatives) obeys the quantum mechanics from which the reduction postulate has been excluded. Therefore this world remains reversible. The irreversibility, which manifests itself in the selection of one alternative or another, appears only as a phenomenon of consciousness. In other words,inanimate matter may be described in EEC in terms of the 4-dimensional space-time in which all time moments are treated equally. The notion of 'the flux of time', of the relations between the present, the past, and the future, together with irreversibility, appears only in the description of the life phenomenon.

Chapter 5

Consciousness and life in parallel worlds: Details for physicists

As it has been shown in Chapter 4, consciousness can be considered to be the *separation of classical alternative realities, or Everett's worlds*. This allows consciousness to "choose" the (subjectively perceived) reality that leads in future to the most advantageous state of the world. For describing this feature of consciousness one needs not only classical realities at the given time moment (results of instantaneous measurements, or Everett's worlds), but also the chains of the realities in subsequent time moments, i.e. the Everett's scenarios. Consideration of this circle of ideas may be more easy or clear in terms of continuous measurements instead of instantaneous measurements. The corresponding mathematical instruments will be very briefly discussed in the present chapter.

This chapter contains the detailed discussion of the classical character of the alternatives and its connection with the phenomenon of life that has already been considered in Sect. 4.3.1. The chapter will be concluded by the proposal to model the quantum concept of life on quantum computers.

The present chapter may be skipped without detriment for understanding of the subsequent chapters. Even those who are interested in mathematical formalism may skip this chapter in the first reading.

5.1 Representation of alternative scenarios by path corridors

Until now, while referring the subject of measurements, we implied instantaneous measurements. That is why the role of alternatives was fulfilled by the state vectors representing the superposition components (in the simplest example that we systematically used, these were $|\psi_1\rangle|\Phi_1\rangle$ and $|\psi_2\rangle|\Phi_2\rangle$). We now consider a more general and more realistic situation where the measurement goes on continuously. In the case of a continuous measurement, the alternative results (readouts) of the measurement may be represented by path corridors.

5.1.1   *Continuous measurements and corridors of paths*

In reality, instantaneous measurements do not exist at all; every measurement has a finite duration. In some cases, the measurement duration is negligible, and it can then be treated as being instantaneous without making a serious error. This is when we are dealing with an instantaneous measurement. Instantaneous measurements are good for analyzing some features of quantum measurements without complicating this analysis by technical details. This is precisely what we have done until now.

In reality, however, measurements most often cannot be treated as being instantaneous: one has to take their duration into account and consider such measurements as continuous ones. In some cases, the duration of a continuous measurement is very long. This is particularly true of the situation where the 'quantum measurement' is not specially organized by the experimenter but emerges spontaneously as a result of uncontrollable interactions of the quantum system with its environment. In this case, the environment is quite frequently termed a reservoir.

The simplest example of a spontaneously occurring continuous measurement is 'quantum diffusion', i.e., the motion of a microscopic particle through some medium. On its way, such a particle permanently interacts with the molecules of the medium that find themselves near the particle. As a result, the state of the molecules changes, so that the information about the location of the particle and its momentum remains in the reservoir, and a measurement (with some finite resolution) of the particle trajectory takes place. Back reaction of the environment (the reservoir) on the particle may be considered as the effect of its measurement by the medium.

Continuous measurement may be represented as a sequence of a large number of instantaneous measurements that occur frequently enough. It may also be described with the aid of the bundles of Feynman paths, which may be visualized as path corridors [Mensky (2000b, 2003)]. A discrete analogue of suchlike corridors are quantum histories [Gell-Mann and Hartle (1993)].

Path corridors play the same part with respect to quantum-mechanical processes as the reduction procedure does with respect to the states of quantum systems. In Feynman's approach, the evolution of a quantum system is described by the integral over all possible paths in the configuration or phase space of this system. When a system undergoes continuous measurement, its evolution is represented by the integral over some corridor of paths. In this case, the corridor of paths itself (denoted by α) corresponds to the measurement result.

5.1.2 *Evolution of a continuously measured system*

The evolution of a continuously measured system during some finite period is thereby subjected to 'projection' in accordance with the result of continuous measurement. This is quite similar to how the state of the system is, in von Neumann's reduction, projected in accordance with the result produced by the instantaneous measurement of this system.

Just as the instantaneous measurement is characterized by the alternative states of the system $\{|\psi_i\rangle|\Phi_i\rangle\}$, so also is continuous measurement characterized by a family of alternatives $\{\alpha\}$, each of which is represented by a path corridor. As with instantaneous measurements, different alternatives are characterized by probabilities, which can be calculated on the basis of quantum mechanics.[1]

For the subsequent discussion, it is significant that each alternative describes semiclassical motion of the system whenever the corridors are wide enough and that the corridor α representing some alternative corresponds to some classical trajectory.[2] At the same time, quantum effects cannot be completely eliminated. This shows up in that the quantum corridors α', α'' coinciding on some interval, may differ as a whole, whereas defining some interval of a classical trajectory completely defines the entire trajectory.

An example of a semiclassical quantum state is the *coherent state* of a family of photons.It is closest to the state of a classical wave.[3] Given the initial conditions, the evolution of the coherent state is well-approximated by the evolution of a classical wave, which is determinate, predictable. An example of an unstable state is the sum or difference of coherent states with

---

[1] The possibility of characterizing the corridors by probabilities (more precisely, by probability densities) instead of amplitudes arises from the fact that they are approximately decoherent, i.e., the interference between them is quite weak [Mensky (2000b, 2003)].

[2] This is true when the corridors $\{\alpha\}$ represent the behavior of not only the system under measurement but also its environment, or the measuring device (just as the alternatives $\{|\psi_i\rangle|\Phi_i\rangle\}$ represent, in the case of an instantaneous measurement, the state of both the system being measured and the device).

[3] The term 'coherent state' refers to the phase of this classical field and is not directly connected with the terms 'quantum-coherent regime', 'decoherence' etc. where the phase factors of the coefficients of a superposition are meant.

strongly different characteristics (the terms may, for instance, correspond to oppositely phased classical waves).

In recent years, suchlike states of a small number of photons have been successfully generated in experiments, and therefore it has been borne out experimentally that they quite rapidly decay with the production of coherent (i.e., close-to-classical) states. The decay occurs due to decoherence, which emerges in the interaction with the environment (from which it is impossible to become completely isolated despite any precautions). Because states of this kind are superpositions of two states close to strongly different classical configurations, *these states have come to be known as the Schrödinger cats* (by analogy with the superposition of a live and dead cat).

Let us underline that the "Schrödinger's cats" that are formed with a number of photons are the products of real experiments, contrary to the Schrödinger's cat in the corresponding thought experiment (see Sect. 1.6.2.1). The radical (althogh not principal) difference is that the Schrödinger's cat is a macroscopic body (containing the nember of the degrees of freedom of the order of 10^{-23}) while the number of photons is a mesoscopic system (of the order of ten degrees of freedom).

Feynman path integrals and integrals over path corridors are mathematically rather complicated (see Refs. [Mensky (2000b, 2003)]). However, we do not need specific calculations here, and do not therefore confront mathematical difficulties. In return, in general reasoning, we can take advantage of the apt illustration of a quantum corridor: the system under measurement moves through a corridor defined by the measurement result. Although a corridor in the phase space is implied in general, we may envision, for clarity, a particle moving in a corridor in our ordinary 3-dimensional space. An alternative in the case of a continuous measurement is the corridor of paths α. And a family of alternatives is a family of corridors $\{\alpha\}$.

5.2   Why alternatives are classical: prerequisite to the existence of life

We have already discussed shortly in Sect. 4.3.1 the important question: why the alternatives separated from each other by consciousness have classical character. Then we have used for this the notion of *"Everett's scenario"* as a chain of the alternative classical states of the world in different time moments. Here, for the readers that are more experienced in quantum physics, we can make use of the notion of the *corridor of paths*.

5.2.1   *Classicality of alternatives corresponds to the experience*

When we consider alternative results of a continuous measurement (alternative path corridors) $\{\alpha\}$ in the framework of the quantum theory of measurement, they should be selected such that they, first, be (approximately) decoherent and, second, (approximately) classical. The *decoherence* is required for the interference between two different alternative evolutions to be weak and the alternatives be characterized by probabilities instead of probability amplitudes [Gell-Mann and Hartle (1993); Mensky (2000b, 2003)]. The *classicality* requirement is not necessary for the absence of interference [Paz and Zurek (1993)] but is introduced in order that the theory correspond to experiment.

Indeed, when conducting any measurements, the experimenter may arrive at different alternative measurement results, but each of these results α, according to his observations, is compatible with the laws of classical physics (the Schrödinger cat may be dead or alive, but not a superposition of the alive cat and the dead cat). For the theory to describe precisely what is experimentally observed, each corridor α must represent a (semi)classical evolution of the system being measured and its environment.

Therefore, the requirement that the alternatives be classical permits constructing the measurement theory that corresponds to observations. But is it possible to theoretically substantiate this requirement? We now see that it is possible if we adopt Extended Everett's Concept, i.e., *identify the separation of alternatives with the consciousness.*

5.2.2   *Classicality of the alternatives from EEC*

If we adopt the Extended Everett's Concept, the separation of alternatives is nothing but the consciousness, i.e., the function inherent only in living creatures. Therefore, the entire set of alternatives, i.e., the definition of what states are considered as alternatives, should be considered bearing in mind that this set is to be used by living creatures. Consequently, we can ask the question: what set of alternatives $\{\alpha\}$ is preferred among all possible sets from the viewpoint of living creatures?

Each alternative α describes the behavior of the whole quantum world.[4] It is described in the same manner as this behavior is perceived by the con-

---

[4]In the context of quantum theory of continuous measurements this may be a microscopic system under measurement and its macroscopic environment, i.e., the whole world. Then the alternatives $\{\alpha\}$ decohere, see [Mensky (2003)].

sciousness. This picture of the world emerges in the consciousness of a living creature. When the world in this picture behaves in accordance with classical laws, it is 'locally predictable' (i.e., the future of some small domain of this world can be predicted with a sufficiently high probability even without knowing what occurs outside this domain). Seeing the predictable world around, a living creature can work out the optimal strategy for survival in this world.

If the alternatives were non-classical, essentially quantum, a picture of an unpredictable world would emerge in the consciousness. In this world, in particular, a significant part might be played by quantum nonlocalities. then the elaboration of the optimal strategy (for a local living creature) would be completely impossible, i.e., life would be impossible (at least the life in the form known to us). The predictability of evolution, which is characteristic of semiclassical corridors $\{\alpha\}$ (which are more correct images of classical trajectories in quantum theory), turns out to be absolutely indispensable in the framework of the Extended Everett's Concept.[5]

Therefore, the classicality of Everett's worlds in the EEC proves to be indispensable to the very existence of living creatures (which may be, with some reservations, considered "conscious" at least at the primitive level, sentient, capable to perceive the environment). As a matter of fact, in the framework of the EEC, quantum mechanics sheds light on the very notion of life, of living matter.

Unlike inanimate matter, a living creature has the ability to perceive the quantum world in a special way. This world, with its characteristic quantum nonlocality, is perceived by a living creature not as a whole but in the form of individual classical projections. Each of these projections is 'locally predictable'. In each of them, the living creature realizes the scenario termed life, while the very notion of life appears to be impossible without this separation.

Therefore, the choice of precisely the classical evolutions α as the alternatives that are separated in the observer's consciousness is favorable for living creatures. This makes plausible that the phenomenon of separation of the alternatives which has been identified with the consciousness (or, for primitive living beings, another way of reflection of the environment, that may be called *"pre-consciousness"*) is not an "absolute" law of nature, but rather *a capacity developed by living creatures in the course of evolution.*

---

[5]This has something in common with the 'existential interpretation' of quantum mechanics proposed by Zurek [Zurek (1998)].

To be more exact, the capacity of pre-consciousness (ability to reflect the state of an environment as the set of separated classical alternatives) had to appear simultaneously with the phenomenon of life. Indeed, it is only after the emergence of this capacity that the quality needed for survival arose and therefore living creatures appeared. However, this formulation may be insufficiently exact, too utilitarian. More likely, the *(pre)consciousness* (= separation of alternatives) is nothing more nor less than. *the definition of what life is* in the most general sense of the word.

If we accept Extended Everett's Concept, we have to conclude that *the classical world does not exist objectively* at all and the illusion of the classical world emerges only in the consciousness of a living creature. Interestingly, we are led to this physically strange conclusion by physics itself, albeit when we bring it to logical completeness to avoid convenient eclecticism like the Copenhagen interpretation with the reduction postulate.

Different attempts to construct the theory of evolution of living creatures in the framework of the Many-Worlds interpretation were made in Refs. [Albert (1992); Chalmers (1996); Deutsch (1997); Donald (1990); Lehner (1997); Lockwood (1989); Penrose (1994); Saunders (1993)].

5.2.3   Modelling of "quantum concept of life" on quantum computers

The picture drawn of the functioning of the consciousness and of its role in the survival of a living creature seems so dissimilar to what we directly see in our classical world that there involuntarily arise doubts as to whether this picture can somehow be verified or is doomed to remain merely a theoretical supposition. As is discussed in Sect. 4.4, this supposition may be confirmed by observations of the consciousness. Here, we remark that direct physical experiments allow verifying at least the fundamental possibility that such 'quantum consciousness' may indeed exist. This requires constructing the model of the 'quantum consciousness' on the basis of a quantum computer.

Indeed, the quantum states evolving in a quantum computer are superpositions with a large number of components. Each superposition component carries some information (e.g., a binary number). The evolution of the entire superposition ensures quantum parallelism, i.e., the simultaneous transformation of all these variants of classical information. In the model of quantum consciousness, individual superposition components can model the classical alternatives into which the consciousness divides the quantum state. The information contained in each component presents then the state of a living creature and its environment.

The problem is to formulate the survival criterion and select the evolution law such that the evolution of every alternative (superposition component) is predictable, so that survival in this evolution be possible. Of course, the task of constructing this model is by no means simple but is basically solvable using a quantum computer.

It is well known that quantum computers, which promise extraordinary new capabilities in mathematical calculations, have not been realized, and some experts doubt that they will be realized in the future (see, e.g., review Ref. [Valiev (2005)]). However, this applies only to quantum computers with the number of binary cells ('qubits') of the order of a thousand or more. As for quantum computers with the number of cells around ten, they have already been realized. Evidently, the number of cells attained will increase further, though maybe slowly.

It is conceivable that even with these 'low-power' quantum computers, which will be constructed in the relatively near future, it will be possible to realize the model of 'quantum consciousness'.

Chapter 6

"Three great problems in physics" according to Vitaly Ginzburg

The Nobel prize winner Vitaly Ginzburg enlisted in his papers (see for example [Ginzburg (1999)]) 30 most important problems of physics and besides this *"the three great problems"* that are interesting from wider viewpoints, including philosophy and concept of life. These "three problems" are 1) interpretation of quantum mechanics, 2) the time arrow, and 3) reductionism (i.e., the question of reducing the phenomenon of life to physics). These actually great problems are being discussed for many decades with permanent interest because they presuppose connections between the areas seemingly far from each other.

We shall discuss these three great problems in the framework of the Extended Everett's Concept (as has been first made in the paper [Mensky (2007a)]). It will be shown that the status of these problems substantially depends on how the first of them is solved, i.e., which interpretation of quantum mechanics is adopted. The Copenhagen interpretation, the Everett's ('many-worlds') interpretation, and at last EEC will be considered.

The viewpoint based upon EEC allows to discover that the "three great problems" are closely connected with each other. Considering them together with each other with the help of the specific assumptions accepted in EEC allows one to take the radical new step in the formulation, understanding and to some extend solving each of these problems.

The material of the present chapter has interest of its own (mostly for physicists, but not only for them). This is why it is exposed to a high degree independently of the other chapters, particularly at the cost of briefly repeating some issues of the preceding chapters (but of course with other points highlighted in these issues). On the other side, this chapter may be skipped without detriment to understanding the next chapters.

6.1 Introduction

At the end of Vitaly Ginzburg's list of the most important problems in physics, we see three problems that are not included on the major list. Listed separately, they are termed by Ginzburg as "the three great problems". These are the interpretation of quantum mechanics, the time arrow (i.e., the irreversibility of time appearing despite the reversibility of the main dynamic equations), and reductionism (i.e., the possibility of reducing the phenomenon of life to physics).

Perhaps these are the most challenging problems faced by physicists and, at the same time, the most interesting ones, or at least the most exciting. Much has been written concerning these problems and certainly many significant results have been achieved. It is definitely impossible to give a complete overview of "the three great problems" here. I will only discuss this subject from a single viewpoint which can be characterized as follows.

I will start the analysis from the first of the three great problems, which is the interpretation of quantum mechanics. I will try to show how the relationships among the three great problems look depending on the way the first one is solved, i.e., in the framework of one interpretation of quantum mechanics or another.

Whenever one speaks about the interpretation of quantum mechanics, the question is always closely related to quantum measurements, since it is the description of measurements of quantum systems that evokes the problem of the interpretation of quantum mechanics. One can therefore reformulate the first great problem as the problem of *quantum measurement theory*. This theory, along with the interpretation of quantum mechanics, is now intensively discussed all over the world, in particular, in connection with quantum informatics. The reason is that the applied field of research called quantum information science is based on the same principles as quantum measurement theory, the principles that are closely related to the interpretation of quantum mechanics. Due to the importance of quantum informatics applications, the last decades have seen a revival of interest in the interpretation of quantum mechanics and the rapid advancement of the relevant field of research.

At the center of modern studies in this field is the interpretation of quantum mechanics suggested by H. Everett in 1957 and often referred to as the 'many-worlds' interpretation [Everett (1957); DeWitt and Graham (1973)]. At the same time, the Copenhagen interpretation, the oldest and the best-verified, has received wide recognition among physicists. It was developed by Niels Bohr in the course of intensive and difficult discussions with other founders of quantum mechanics, in particular, with Albert Einstein. These two interpretations are qualitatively different, while other numerous interpretations are just versions of these two and differ from them only in details.

The present author suggested in 2000 [Mensky (2000a)] a generalization of the Everett's interpretation called the Extended Everett's Concept (EEC).[1] In contrast to the original interpretation by Everett, EEC leads to new predictions that relate to the work of consciousness.

Below, in the analysis of "the three great problems" I will rely on (1) the Copenhagen interpretation, (2) Everett's interpretation, and (3) the EEC. In fact, these concepts are the three ways to solve the first of "the three great problems". After briefly characterizing each of the three different approaches to the conceptual problems of quantum mechanics, I will try to trace how the remaining two of "the three great problems" and the relationships among all three look in view of different approaches.

In other words, interpretation of quantum mechanics will be the starting point for the discussion of, first, the phenomenon of life (and the question of whether it can be explained in the framework of quantum physics) and, second, the 'time arrow' problem (i.e., the question why, despite the reversibility of quantum-mechanical evolution, there is still irreversibility in quantum mechanics).

6.2 'Ginzburg's problems'

The well-known Russian physicist, 2003 year Nobel prize winner, Vitaly Ginzburg was always known by his strong interest to foundations, history and perspectives of physics and science generally. This permanent interest allowed him to create the widely known "Ginzburg's seminar", which attracted hundreds of physicists each Wednesday during 45 years.[2] The same permanent intensive scientific curiosity together with the deep and systematic approach to science reveals itself in his work on the list of the most important problems in physics (see [Ginzburg (1999)] for one of the last publication of this list).

---

[1]It is considered in detail in Chapters 4, 5.

[2]The author had the honor to give talks at this seminar. After some of these talks V.L.Ginzburg invited me to write papers for Physics-Uspekhi. One of the review talks and the subsequent papers [Mensky (2000a, 2005a)] laid the foundation of the approach to the theory of consciousness that is exposed in the present book.

In addition to solving specific problems, Ginzburg has constantly analyzed physics as a whole, as well as science as a whole, and has always looked for the points of growth in physics and in science in general. This is certainly very important since the future of science grows from current problems. A today's problem may tomorrow become the center and essence of all physics and a source of new achievements. In any case, this is true for "the great problems", against which scientists have struggled for many decades, not losing interest but, on the other hand, not considering the achieved progress as a final solution.

The "three great problems", which Ginzburg mentions at the end of his list, can be formulated as the following questions:

- Interpretation of quantum mechanics: what happens during measurement?
- The phenomenon of life and reductionism: what is life from the viewpoint of physics?
- The time arrow: where does irreversibility come from?

The first problem is called the problem of the interpretation of quantum mechanics but in fact is an attempt to find out what happens during measurement. Why during measurement? Because conceptual problems (paradoxes) of quantum mechanics present themselves when we try to analyze the process called measurement in terms of quantum mechanics. In classical physics, the description of measurement is very simple, in fact, trivial (of course, only in principle, when purely technical issues related to the measurement devices are ignored). However, it turns out that the quantum-mechanical description of measurement causes paradoxes. It is not at all evident what 'measuring a quantum system' means and what happens in such a measurement.

The second problem is the phenomenon of life and reductionism. What is life from the viewpoint of physics? Can one explain the phenomenon of life based on the laws of physics? There is no evident answer to this question. In any case, the numerous attempts to 'derive the phenomenon of life from physics' (together with other natural sciences) have enjoyed no success so far.

The third great problem, according to Ginzburg, is the origin of the time arrow. Where does irreversibility come from? In quantum mechanics, which is the most fundamental science, all equations are reversible in time. How, then, does irreversibility appear?

Let us start the analysis by choosing one interpretation of quantum mechanics or another and discuss the other two problems and especially the relationship among all three problems from the viewpoint of the chosen interpretation.

In this connection, it is important to note that various interpretations of quantum mechanics are, in fact, various levels of describing quantum measurement. Sometimes one says: let us find which interpretation is the correct one. In my opinion, this is the wrong question. Various interpretations are different descriptions of the same process, the quantum measurement. All descriptions are correct but they 'decode' this process on various levels, uncovering the mechanism of measurement to a greater or lesser extent.

Starting from the famous paper by Einstein, Podolsky, and Rosen [Einstein, Podolsky and Rosen (1935)], it has become more and more evident that to give an interpretation of quantum mechanics means to explain *how reality is understood in quantum mechanics* or, in other words, what the *quantum reality* is. Accepting one interpretation or another means explaining the quantum reality in one way or another. This, however, can be done on various levels.

More primitive levels (including the Copenhagen interpretation) are rather easy for understanding and convenient in practice, but their description of the essence of quantum reality is not sufficiently exact. Interpretations of higher level (including Everett's) express this essence more precisely but are more arduous for understanding and produce more difficulties than help for practicing researcher in quantum physics (for instance, for solving typical quantum-mechanical problems).

This explains why Everett's interpretation was accepted with such difficulty. Nevertheless, it has come into great demand over the last decades, in particular, due to the development of quantum information science.

A more exact (of higher level) interpretation does not cancel a less exact one, since different interpretations do not influence the mathematical base of quantum mechanics. Calculations and predictions for specific experiments are made according to the same recipes regardless of the interpretation. As a consequence, calculations do not require complicated interpretations like Everett's one. The Copenhagen interpretation is quite sufficient. But if one takes into account the fact that the Copenhagen interpretation has logical defects, then, for more exact reasoning, one has to turn to other interpretations and, first of all, to Everett's.

It is important however that moving to more profound interpretations of quantum mechanics not only restores the logical completeness of this

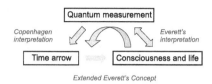

Fig. 6.1   Description of the quantum measurement on a higher level reveals a richer structure of relations between the three great problems.

science but also allows one to explain important facts relating to the work of the consciousness (i.e., at first sight, having no relation to quantum physics) that have found no explanation up to now (see Chapter 4 and Sects. 6.7, 6.8).

6.3   Relations among "the three great problems"

Let us begin with some reasoning leaping ahead. We will briefly characterize the relationships among the three great problems without dwelling on the proofs of these relationships. This reasoning can be illustrated with the scheme shown in Fig. 6.1.

Quantum measurement and the time arrow — what is the relation between them? It is very simple. This relation was discussed long ago, it is evident even in the framework of the *Copenhagen interpretation* of quantum mechanics in which the description of measurement by the state reduction is accepted.

According to the Copenhagen interpretation, at the moment when an (instantaneous) measurement is performed in a quantum system, some abrupt change (jump) of its state occurs, and this change is *irreversible*, contrary to the reversible evolution of a quantum system due to the Schrödinger equation. This change is called the *state reduction* or the wave function collapse. Such an irreversible change in the state resulting from the measurement occurs only in quantum physics; a measurement in classical physics does not lead to irreversibility (although irreversibility may emerge in classical physics by other reasons).[3]

Before the measurement, only the probabilities of various measurement results can be predicted, even if the state of the system is fully known. During the measurement, a single result is chosen from the set of all pos-

---

[3] We consider here only the irreversibility connected with quantum measurements. Some classical system, such as the Sinai billiard, show irreversibility, but of a completely different origin, namely, due to the instability with respect to initial conditions.

sible (alternative) measurement results. Then the state of the system is irreversibly changed into the state whidh is compatible with the given measurement result. After the measurement, the system cannot return to the initial state in which all measurement outcomes are possible. This way, a measurement brings irreversibility to quantum mechanics, which is absent in a usual (Schrödinger's) quantum evolution, with no measurements involved.

This reasoning is valid in the framework of the Copenhagen interpretation of quantum mechanics. However, if we consider a more complicated interpretation, *Everett's interpretation* (we shall see later why the Copenhagen interpretation is not sufficient), then it turns out that the observer's consciousness should be included in the measurement. Without this inclusion, the measurement description is not complete. Thus, new relationships among the three problems appear.

First of all, a relation appears between quantum measurement and consciousness. This relation is hardly expected from the usual physics viewpoint. Indeed, consciousness is a phenomenon of living creatures. This notion is simply absent in the world of inanimate physical systems that is the subject of physics. By introducing the consciousness of the observer into the measurement theory, Everett's interpretation of quantum mechanics establishes a direct relation between quantum measurement (and, hence, quantum mechanics in general) and the phenomenon of life. This seems to be completely foreign to physics, at least in its simple version, which one could expect to describe measurement.

Therefore, it turns out (in the framework of the Everett's interpretation, that quantum mechanics and the phenomenon of life are closely related and the measurement theory in quantum mechanics proves to be not so simple.

In addition, in Everett's interpretation the irreversibility of measurement arises only in the picture drawn by the observer's consciousness. Hence, measurement leads to the time arrow only if the role of consciousness is taken into account (see the bent arrow in Fig. 6.1).

If we pass further to the Extended Everett Concept (it will be discussed later why this extension is necessary), then another arrow appears, another relation among the above three problems — connection of the time arrow with consciousness and life.

Indeed, in the framework of the EEC one can understand how a decrease in entropy can occur in life, while the rule for inanimate systems is the entropy increase. For life, self-organization is typical. Life develops, and its evolution is directed not toward greater chaos (an entropy increase) but

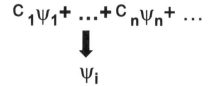

Fig. 6.2 Reduction postulate: when a quantum system is measured, its initial state changes in such a way that a single component of the superposition survives, the one that corresponds to the measurement result.

toward greater order (an entropy decrease). In the living world, the time arrow also exists, but the entropy behavior with respect to the time arrow is strange: the entropy decreases.

This is caused by the fact that the very existence of a living creature depends on what will happen in the future. The notion of *goal* (the basic goal is survival) is inherent in living world and hence the related feeling of *time running*: the future is different from the past and present (which separates the future from the past).

Summarizing (and leaping ahead, since we have not proved anything so far), we see that solving the first problem leads to a deeper understanding of the other two, according to the following scheme:

- Quantum measurement
  ⇒ the role of the observer's consciousness
⇒ • Phenomenon of consciousness and phenomenon of life
  ⇒ quantum reality
⇒ • Reversibility of the quantum world and the subjective feeling of time running.

6.4 Copenhagen interpretation: state reduction

What is the Copenhagen interpretation? How does it describe a measurement in a quantum system? Briefly, this can be formulated in the following way (Fig. 6.2). Let the state of a quantum system before the measurement be the superposition state $c_1\psi_1 + \cdots + c_n\psi_n + \ldots$, where the components $\{c_n\psi_n\}$ correspond to various measurement results that can be obtained with a given instrument. Then, after the measurement the system is brought into a single definite state ψ_i, one of those forming the superposition. This effect, i.e., *selection of one of the components and the disappearance of all other ones*, is called reduction of the state, or the wave

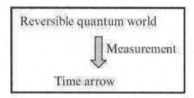

Fig. 6.3   The reduction postulate means that a measurement of a quantum system leads to an irreversible change in its state, i.e., a quantum measurement leads to irreversibility (creates the time arrow).

function collapse. This change in the state of a quantum system is stepwise and irreversible. A measurement causes a jump from a superposition state into a state given by a single component of the superposition.

Thus, in the framework of the Copenhagen interpretation a measurement of a quantum system is an irreversible process. Measurements (and not only ones that are specially organized but also those that occur spontaneously due to the environment or a thermostat) introduce irreversibility into quantum mechanics. Hence, the first relation between the great problems is established: *measurement brings the time arrow into quantum mechanics.* In a theory, whose equations are symmetric with respect to the time reversal, irreversibility appears (Fig. 6.3). This kind of irreversibility has been intensively discussed in the literature, a review can be found in the monograph [Zeh (1992)].

In the Copenhagen interpretation, the state reduction was postulated. A mathematically strict formulation of this postulate was first given by von Neumann, and therefore the reduction postulate is also called the *von Neumann postulate.* If one postulates that a measurement causes reduction, i.e., one of the superposition components is *selected* (with a corresponding probability), then calculations based on this postulate will never lead to errors.[4] In this sense, such a postulate makes quantum mechanics efficient.

If the reduction postulate is rejected, a problem appears. We understand very well how a closed system behaves, but if such a system is measured and still exists after the measurement, then a question arises: what is the state of the system after the measurement? Indeed, to find out what happens with the system later (during a time interval after the measurement), we need to know its state immediately after the measurement.

---

[4]Of course, a more realistic description of quantum measurements requires some purely technical generalizations, in particular, the concepts of a soft (inaccurate) measurement and a continuous measurement (see, for instance, Ref. [Mensky (2000b)]), but these refinements do not change the essence of the questions we are discussing.

If this question is answered the same way as in the reduction postulate, then calculations provide predictions that are confirmed experimentally. In this sense, the reduction postulate leaves no doubts. The genius of N. Bohr allowed him, in particular, to develop such a simple formulation of quantum mechanics (the Copenhagen interpretation) that could efficiently solve quantum-mechanical problems despite the conceptual gaps that, as many researchers understood, still remained in this science.

6.5   Measurement as an interaction: entanglement and decoherence

However, the reduction postulate itself can be doubted. And it was doubted from the very beginning, but these doubts became more solid after the development of the decoherence theory. The decoherence theory describes the measurement process without invoking any special postulate like the reduction one, and only within the framework of the conventional quantum-mechanical formalism where evolution is described by the time-reversible Schrödinger equation.

To move to this description of measurement, it is sufficient to recall that a measurement is the interaction of the measured system with another system, which can be called the measuring instrument. This second system can be considered to be the environment of the system under measurement. Interaction of the system under measurement with its environment can be described in the framework of standard quantum mechanics based on the Schrödinger equation and not involving the reduction postulate. In this case, one should try, in the framework of standard quantum mechanics and without the reduction postulate, to answer the question: what happens when a measurement is performed on a quantum system?

This turns out to be possible. An appropriate consideration shows that in the process of interaction of the measured system with the measuring device the states of these two systems become *entangled*, or *quantum correlation* between them appears. Further analysis shows that *the state of the measured system, taken separately* from the measuring device, is crucially changed as a result of the measurement (i.e., its interaction with the environment). Instead of the *pure state* as it has been before the measurement, the measured system is now in a *mixed state*. Instead of the sum (superposition) of the vectors, or the wave function components (corresponding to the alternative results of the measurement), the mixture of the same vectors appears after the measurement. One says that in this case the measured system undergoes *decoherence*.

Why is this change in the state of the measured system called decoherence? Because the system subjected to measurement (interacting with a measuring instrument) loses quantum coherence. The information about the relative phases of the components in the superposition (wave function components) is lost after the measurement. As a result, if the state of the system before the measurement was a pure one and was described by a wave function (a state vector), after the measurement it becomes mixed and is described by a density matrix.

Most probably, the physical essence of measurement was already clear to the founders of quantum mechanics, but at that time it was neither stressed nor formulated in detail. Therefore, it was much later that this science, the decoherence theory, was rediscovered by the scientific community. The phenomenon called decoherence became widely known starting from 1982, when the paper [Zurek (1982)] by W. H. Zurek appeared. After that, decoherence was extensively discussed in the literature, and its understanding gradually deepened.

It turned out that physicists have been constantly facing this phenomenon while studying various systems and their interactions, but it was not considered to be a special class of quantum-mechanical processes. Zeh and then Zurek described this class of processes from the viewpoint of quantum measurement theory and thus revealed its special role. After that, physicists started to actively study the decoherence effect.

In the course of these studies, it developed that decoherence had been understood and very well-described as early as 1970 by Dieter Zeh, a German physicist [Zeh (1970)] (see also [Joos and Zeh (1985)]). However, neither the paper [Zeh (1970)] nor later works by Zeh and his disciples were noted by the scientific community. In 1979, decoherence was described by the author of the present book in the framework of a completely different phenomenological approach, based on Feynman's path integrals [Mensky (1979a,b)]. Still, it was only many years later that various ways of describing this process were brought together and compared, and all works were understood as relating to the same class of phenomena which was called decoherence. The very term 'decoherence' was introduced in a paper by M. Gell-Mann and J. B. Hartle [Gell-Mann and Hartle (1990)] only in 1990. The modern state of the decoherence theory is well-described in the frameworks of various models in the book [Giulini et al. (1996)] by Zeh and his students, while its description from the viewpoints of different phenomenological approaches can be found in the book [Mensky (2000b)] (in these approaches, the environment is not considered explicitly and its influence on the system is taken into account phenomenologically.)

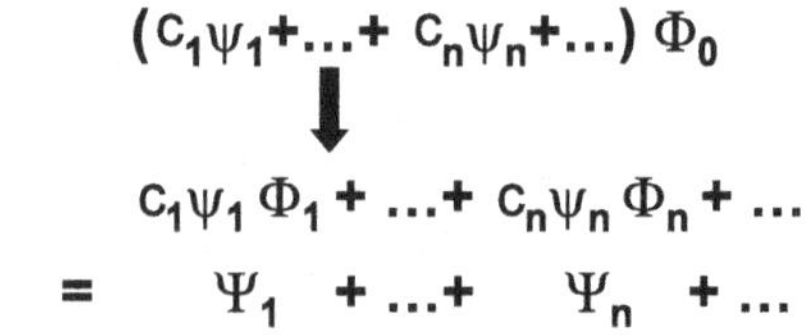

$$(c_1\psi_1 + \ldots + c_n\psi_n + \ldots)\,\Phi_0$$

$$\downarrow$$

$$c_1\psi_1\,\Phi_1 + \ldots + c_n\psi_n\,\Phi_n + \ldots$$

$$=\quad \Psi_1 + \ldots + \Psi_n + \ldots$$

Fig. 6.4   Due to the linearity of quantum mechanics, the state reduction is impossible. During a measurement there occurs only 'entanglement', or quantum correlation, between the measured system and the instrument, leading to the decoherence of the measured system.

Let us discuss in more detail what happens during a measurement of a quantum system. How can one consider measurement in the framework of conventional quantum mechanics? This is illustrated in Fig. 6.4. Similarly to our previous reasoning, we assume that before the measurement the system resides in a superposition state $c_1\psi_1 + \cdots + c_n\psi_n + \ldots$, but now we will take into account not only the system under measurement, but also the measuring instrument, or the environment of the system. Let the state of the environment before its interaction with the system (i.e., before the measurement) be described by the vector Φ_0. Then, the state of the whole system, whose subsystems are both the measured system and the environment, before the measurement is given by the vector $(c_1\psi_1 + \cdots + c_n\psi_n + \ldots)\Phi_0$. Now, let us consider the interaction between the measured system and its environment and ask the following question: what happens after the interaction? How do the states of the system and its environment change? It turns out that under some natural assumptions about the interaction, the conventional quantum mechanics makes the state of the total system change in the following way:

$$(c_1\psi_1 + \cdots + c_n\psi_n + \ldots)\Phi_0$$
$$\to\ c_1\psi_1\Phi_1 + \cdots + c_n\psi_n\Phi_n + \ldots$$

Here, Φ_n denotes the state of the measuring instrument that is interpreted by the experimentalist[5] as indicating the system to be in the state ψ_n.

---

[5]The description of the environment is maximally idealized here, but without loss of any significant features of the process. In reality, the Φ_0 vector represents only some part of the environment, the one that directly interacts with the system under measurement; this part of the measuring instrument is usually called a meter. In the general case, before the interaction (measurement) it can be in any one of the set of states $\Phi_0^{(\lambda)}$ (which become, respectively, $\Phi_n^{(\lambda)}$ for the nth measurement result), or in a mixed state $\sum_\lambda p_\lambda\,\Phi_0^{(\lambda)}\Phi_0^{(\lambda)\dagger}$.

We now see that the state vector of the system under measurement and the state vector of the measuring instrument (environment) do not exist separately. Instead, there is only the state of the total system, in which the measured system and the measuring instrument are correlated. This 'non-factorisable' state (which cannot be factored into a product of the state vector of the system and the state vector of the instrument) is called *entangled*. In such a state, there is a quantum correlation between the system and the instrument.

The correlation can be formulated in the conditional mood: if the system resides in the state ψ_n (in the nth component of the superposition), then the instrument is in the state Φ_n. However, one should realize that this conventional phrase does not reflect the specific features of quantum correlation, distinguishing it from correlations feasible in classical systems.

It is important for us that in this description all components that were present in the superposition before the measurement are still retained after the measurement (although each component changed). The disappearance of all components except one, which was to occur according to the reduction postulate, did not happen here.

Thus, the usual quantum-mechanical treatment of the measurement event shows that all superposition components survive the measurement. All that happens is the phenomenon termed 'entanglement', or quantum correlation, between the system under measurement and the instrument (the environment).[6]

It is important — and later we shall discuss this from another viewpoint — that the total system, i.e., the measured system and its environment, resides in a superposition state after the measurement. In Fig. 6.4, this circumstance is highlighted in the bottom line: it is not the structure of each component in the superposition that is important but the fact that all superposition components 'survive' the measurement.

Let us summarize our reasoning where measurement is considered as interaction. If a superposition exists at some stage, it will be further retained,

---

[6]Sometimes, one tries to justify the reduction postulate by claiming that the measuring instrument is macroscopic and its evolution is classical. However, the classical description of any system is approximate (compared to the quantum one) and in no way cancels the exact description in the framework of quantum mechanics. (It only makes the quantum approach too detailed when only a crude description of a system is necessary.) Therefore, a conclusion made in the framework of an exact description cannot be disproved by means of an approximate description.

and this follows from the linearity of quantum-mechanical evolution.[7] Each term in the superposition may change somehow, but all terms will still be present, none of them becoming zero. There is no reduction, i.e., selection of a single component and the disappearance of the other ones. This is dictated by quantum mechanics. Quantum mechanics excludes reduction.

6.6   Everett's ('many-worlds') interpretation: no reduction

Thus, if we trust quantum mechanics, i.e., consider the evolution of the system to be always described by the Schrödinger equation, then the evolution in the course of measurement must be presented by entanglement and decoherence rather than state reduction. Reduction postulate should be then somehow excluded. How can one do it? The answer was given by an interpretation of quantum mechanics proposed by H. Everett [Everett (1957); DeWitt and Graham (1973)] in 1957.

The logic upholding Everett's interpretation is very simple. Let us start from quantum mechanics. Quantum mechanics dictates that no reduction is possible. Relying on quantum mechanics, we accept the statement: there is no reduction, all components of a quantum superposition survive during the evolution, including the measurement process (see the bottom line in Fig. 6.4).

However, if one accepts this simple logic, it is necessary to explain how it happens that the observer sees just a single measurement result corresponding to just a single component of the superposition.

A measurement can lead to different results, which exclude each other in the *consciousness of the observer* — the 'alternatives'. All alternatives are still present in the superposition, and Everett's interpretation assumes that after the measurement they are still kept in the description of the state.

How can one understand this? How can one combine this with the everyday experience of an experimentalist who always observes just a single measurement result and not a superposition of results, only a single alternative Ψ_i and not a superposition $\sum_n \Psi_n$ of alternatives? Does not this everyday experience contradict to the assumption about coexisting all the alternatives (as in the bottom line of the scheme in Fig. 6.4)?

---

[7]Retaining of all components of the superposition is provided by linearity and one more property of the evolution, its unitarity, but in the special case of an (ideal) measurement unitarity follows from the requirements to the interaction of the measured and measuring systems.

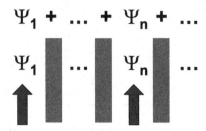

Fig. 6.5 Everett's interpretation: reduction (disappearance of all alternatives but one) does not happen but consciousness separates classical alternatives by perceiving them separately.

It should be noted that now an important role in the reasoning is played by the observer or, more precisely, the *observer's consciousness*, and the interpretation of the fact that all superposition components are retained can involve this notion, the observer's consciousness, his subjective perception of the world.

It is important that the picture subjectively perceived by the observer (which is represented by the Ψ_i vector) is a purely classical one, while the different alternative pictures of the wold (say, $\Psi_{i'}$ and $\Psi_{i''}$) are classically distinct.

Thus the alternatives presented by the components of the superposition $\{\Psi_i\}$, are alternative pictures of the classical world, and it is always only a single one among the alternative pictures of the classical world that is perceived subjectively. (In terms of a measurement procedure, different pictures of the classical world correspond to different positions of the measuring instrument pointer, and the observer always sees just a single position of the pointer.)

In Everett's interpretation, one should explain how this can agree with the fact that the superposition contains all alternatives $\{\Psi_n\}$ corresponding to various pictures of the classical world so that these alternatives are assumed to coexist.

To overcome this controversy, the following statement may be assumed in the Everett's interpretation. All superposition components exist and describe different alternatives, i.e., alternative measurement results or alternative classical (quasiclassical) states of the quantum world, but *consciousness separates the alternatives* Consciousness is the perception of all of these alternatives, but they are perceived separately from each other. If a person is aware of observing one of the alternatives, he cannot be aware of seeing the other ones at the same time.

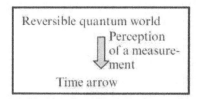

Fig. 6.6   According to Everett's interpretation, irreversibility appears in a quantum measurement due to the *awareness* of a measurement result.

Separation of the alternatives by the consciousness is a formulation of Everett's interpretation that is convenient for our purposes. There are also other formulations, for instance, the one where different classical worlds exist, *Everett's worlds*, which correspond to all possible alternatives. According to this formulation, each of the observers exists in each of Everett's worlds (in other words, an observer has twins, or replicas, in each of the Everett's worlds).

This formulation is very widely spread because of its explicitness, but actually it sometimes causes misunderstanding since it contains a certain inaccuracy: one should speak not of different classical worlds but of different classical states of a single world and about the superposition of these states.

If we accept the statement about the alternatives being separated by the consciousness, then in the description of the the observer's subjective perceiving the same effect occurs as predicted by the reduction postulate: *subjectively, the observer will see (recognize) only one of the alternative classical pictures of the world.* However, now we have managed to combine it with linear quantum mechanics: all alternatives exist in reality (because of quantum character of this reality) but they are separated in the consciousness. Consciousness, similarly to the state of the material world, also consists of something like multiple components, which subjectively seem to be mutually exclusive. These components reflect the alternatives.

What new results does it provide for the relations among "the three great problems"? How do these relations change if one moves to Everett's interpretation in which all alternatives are assumed to be equally real but separated in the consciousness? The relationships among the three great problems remain almost the same as in the case of the Copenhagen interpretation, with the difference in a single nuance (Fig. 6.6). Now, one should say that the time arrow does not objectively exist in the quantum world but it only appears in the consciousness of the observer.

In reality, i.e., in the objectively existing world, all superposition components, all alternatives, are retained (stay equally real), and the evolution of their superposition is quite reversible. However, consciousness perceives these alternatives separately, and this leads to a picture of an irreversible process in the subjective perception. Namely, subjectively the observer perceive the choice of a single alternative and the disappearance of the others.[8]

An observer seeing one of the alternatives does not see the other ones. Subjectively, this does not differ from the picture where one of the alternatives is selected and the others disappear, i.e., from the state reduction picture. However, now, in view of Everett's interpretation, one has to conclude that *the state reduction is just an illusion appearing in the consciousness* of an observer; in other words, it is a specific feature of the consciousness. Therefore, illusion is that the real evolution is irreversible.

6.7 Extended Everett Concept (EEC): definition of consciousness

Let us now move to the Extended Everett Concept (EEC) which allows one to consider quantum measurements at yet a higher level and leads to a number of very interesting consequences [Mensky (2000a, 2005a, 2004, 2005b, 2007b)]. The step that takes us beyond Everett's concept is *identifying consciousness with the separation of the alternatives*. Let us explain this.

6.7.1 *Identity of consciousness and alternative separation*

Let us start from Everett's concept in the formulation used above in Section 6.6: all alternatives exist (there is no reduction) but consciousness separates them. By thinking a little deeper, one can see that, in fact, the two central notions of this formulation are not defined and cannot be defined at present.

Using the notion of 'separating the alternatives', we actually do not fully understand what it means and have to accept just a vague intuitive idea of its meaning. Similarly, while operating the notion of 'consciousness', we do not actually understand what consciousness is.

---

[8]One should not think that in this way one of the alternatives is singled out, namely, the one seen by the observer. He observes (his consciousness perceives) all the alternatives, but he sees them separately.

Physicists cannot explain the separation of alternatives in the framework of quantum mechanics (and, hence, cannot fully clarify this notion), nor can psychologists, physiologists, and philosophers, who actively work on the problem of consciousness inwardness, solve this problem.[9] Apparently, the phenomenon of consciousness is somehow related to the work of the brain, but it cannot be fully explained by the brain functioning.

The Extended Everett Concept (EEC) suggests identification of these two poorly defined notions, the 'consciousness' and the 'alternative separation'. It is assumed that consciousness is identified as the separation of alternatives. After this identification, first, there remains just one notion instead of two and, second, this notion can now be illustrated from two viewpoints: the physical one, and the psychological one. The separation of alternatives, not very clear in physics, is illustrated by what we know about consciousness, while consciousness, which is not very clear in psychology, gets illustrated due to what physics knows about the separation of alternatives.

In fact, one cannot expect more than that. In any science, initial notions stay vague until it becomes clear how these notions work and how all other notions arising in the theory are related to each other. By making the notion consciousness = alternative separation common to quantum physics and psychology, we take a step toward its more exact definition.

Thus, the assumption of identity of consciousness and alternative separation, arbitrary as it may seem at the first glance, is plausible due to the resulting simplification of the logical structure of the quantum theory (in the Everett's "many-worlds" version). Of no less importance (and maybe even more convincing) is the fact that combining these two notions leads to the explanation of some phenomena that are well known but up to now not explained. Particularly, natural explanation is done to the phenomenon of highly efficient intuition often observed, particularly in science.

6.7.2   Consequences of the identification

Both Everett's interpretation and the EEC give, on the one hand, a description of the quantum world represented by a superposition of alternatives and, on the other hand, a description of the same world as perceived by the

---

[9]It is important to underline that in this argument we make use of the term "consciousness" in its most narrow sense that may be better expressed by the words "root of consciousness". We do not deal with many intellectual processes performing in the state of being conscious that are often also denoted by the same word "consciousness".

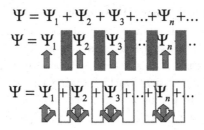

Fig. 6.7   If consciousness and separation of the alternatives are identified, then dimmed consciousness (in particular, in the state of sleep or trance) means an incomplete separation of alternatives, in which consciousness looks into 'other alternatives' and can single out the most favorable ones among them.

consciousness. This is the same quantum world but with separated alternatives. Alternatives constitute different 'projections' of the quantum world. If evolution is described in the framework of quantum mechanics, all these projections are essential and are present together (as a superposition). In the description of the picture existing in the consciousness, the alternatives are separated, and each of them has a meaning but the sum is meaningless. In the Everett's interpretation we say: consciousness separates the alternatives. But in the framework of the EEC we say it slightly differently: it is the separation of alternatives that is consciousness.

At first sight, this identification seems to change nothing essential in the measurement picture. But this is not so. Now, after identifying consciousness with the separation of alternatives, one can pose the following question, which in fact does not relate to physics any more but is outside of its scope: *what happens when consciousness is turned off?* Indeed, states of turned-off or dimmed consciousness are known, these being *sleep, trance, meditation,* or what Young called the *unconscious.* What happens in transferring to such states from the viewpoint of the concept we consider?

Physics cannot answer this question but if we assume that separation of alternatives is identified as consciousness, then the answer is possible. Under this identification, *turning off consciousness means turning off the separation of alternatives.* It is logical to conclude: when consciousness becomes dimmed, the separation of alternatives becomes incomplete, 'partitions' between alternatives become transparent (Fig. 6.7). Immediately, an important conclusion follows: if consciousness is dimmed or weakened, then, while perceiving some alternative, it at the same time scans the neighboring alternative, and not only the neighboring one. Hence, a subject in the state of dimmed consciousness, perceiving some classical alternative, can at the same time look into 'other alternatives'.

This gives qualitatively new ability to a subject. In the state of unconscious he has access to all possible classical states of the world, or, in other words, to all parallely existing Everett's worlds. The information that becomes available because of this access is in principle not available in the framework of a single classical world. Taking (in some form or another) this information and returning to the conscious state, the fellow may take the answer to the question that cannot in principle be answered if think about it in conscious state. This may explain examples of astonishing super-intuition and particularly great scientific insights.

To this must be added the assumption that a subject observing some alternative (while separating them) can *modify the probability of observing one alternative or another in the nearest future.* In the framework of the EEC, this assumption becomes natural because separation of alternatives, after identifying it with consciousness, can be considered in two ways: as a specific description of what happens in the quantum world, and as a mental phenomenon. The quantum world is based on objective laws but mentality is subjective; it is controlled, at least partly, by the subject.

Therefore, it is natural to define two probability distributions in the set of alternatives: the objective probabilities (regulating the choice of an alternative in the world of inanimated physical systems), and the subjective probabilities (defining which alternative will be "chosen" by the subject to perceive it subjectively).

The assumptions of the EEC are quite counter-intuitive and not typical for physics. However, analysis shows that the logical structure of the theory is simpler under these assumptions than in the Copenhagen interpretation or in Everett's interpretation in its original form. But most important is that with these assumptions we become able to explain many things that we face every day but that have had no explanations up to now.

For instance, the *free will.* What is free will? A person wants to leave the room and leaves it, or he wants to stay there and stays. He wants to get up from a chair and gets up, or stays seated if he wants. It seems simple but do we understand how it happens? How is the decision made?

We will not find the answer by analyzing the work of the brain. The command to muscles comes from the brain but how is one of several alternative commands chosen by the neuron that first makes this choice? Physiology cannot explain this. The assumptions adopted in the EEC explain this in a natural way: all alternative behavior scenarios are present as superposition components but the subject can compare them with each other and increase the subjective probabilities for the alternatives that seem more attractive to him (for instance, those more favorable for life).[10]

In addition to the free will, this reasoning can explain such a strange fact as the *absolute necessity of sleep.* Everyone is so used to the phenomenon of sleep that we never think about this fact. But biologists and physicians cannot explain why sleep is absolutely necessary, why a person deprived of sleep for three weeks will certainly die. The answer that sleep gives rest to the organism does not actually explain this absolute necessity. The Extended Everett's Concept explains this phenomenon: a person deprived of sleep has no opportunity to look into 'other alternatives' and choose the best one, leading to maintaining health and survival.

Beside these, there are other fundamental phenomena that find natural explanations in the framework of EEC. Among them, there are also phenomena, probably existing in reality, consisting in observing events that naturally occur only with extremely small probabilities ('probabilistic miracles').

6.8   Extended Everett Concept (EEC): relations between "three problems"

If one accepts the Extended Everett Concept, i.e., identifies consciousness with the separation of alternatives, then the relations between "the three great problems" are again slightly modified, and in this case they become especially diverse. These relations are represented in Fig. 6.8.

(1) According to the EEC, there is a field where only 'pure' quantum theory operates. In this theory, evolution is always described by a linear law (for instance, the Schrödinger equation) and is reversible. The reversible quantum world is represented by the world of inanimate matter. No notion of measurement is necessary in this world: measurement is only the interaction of the system with its environment, and all interactions in the reversible quantum world are correctly described by the usual linear quantum-mechanical equations in terms of the notions of entanglement.

The existence of consciousness, or separation of alternatives, first of all, enables one to explain the phenomenon of life. The key role here is played by the classical nature of the alternatives. By identifying the separation of

---

[10]Of course, if only the phenomenon of free will is considered and the postulates of the EEC are used only for its explanation, then these postulates seem quite voluntary. However, since they originate from a reasoning that starts from quantum physics, the whole construction becomes plausible.

alternatives with consciousness, i.e., with some attribute of living matter, the EEC explains the classical nature of the alternatives, which cannot be explained otherwise.

Indeed, separation of alternatives is consciousness, i.e., an attribute of living matter. Therefore, it is legitimate to pose the question: to what components the quantum state of the world will be separated, and what will be the alternatives (the superposition components) in the interests of life?

The answer is obvious: the alternatives should be classical (quasiclassical), so that consciousness (in the regime of separation of the alternatives) perceives the picture of a locally predictable world (i.e., such a world in which the evolution of some spatial domain cannot substantially depend on the states of remote domains). If, instead of classical alternatives, essentially non-classical ones were used (involving the features of quantum nonlocality), then each such alternative would give a picture of an unpredictable world in which the strategy of survival could not be worked out.

Only classical alternatives provide the predictability of the world sensed subjectively, and hence ensure the very possibility of life.

Further, if one takes into account that consciousness can be in the 'boundary state', in which it is almost completely turned off, i.e., the alternatives are not completely separated, it becomes possible to explain how life is maintained and the health of a living creature is preserved. Here, the main role is played by sleep, during which the dimmed consciousness penetrates into 'other classical realities' (other Everett's worlds), the subject compares alternatives and is enabled to choose the one that is most favorable for life and health. Sleep is absolutely necessary for life namely due to the fact that it helps to choose the strategy for survival. Maintaining life is impossible without sleep.[11]

(2) The second line of relations between "the three great problems" connects the problem of measurement and the problem of the time arrow. Considering consciousness, or separation of alternatives, we necessarily come to the conclusion that the picture created by the consciousness contains something that is absent in the quantum world. The quantum

---

[11] The phenomenon of sleep (periodically turning off clear consciousness and getting into a state of 'being unconscious') exists not only for humans beings but also for animals whose physiology is close to that of human beings. For more simple organisms, 'consciousness' of the same type may be absent, the ability to perceive (or rather reflect) the surrounding world, is probably similar to what is called 'unconscious' for a human being; hence, the phenomenon of sleep is of no significance to them.

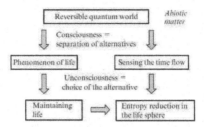

Fig. 6.8 In the framework of the extended Everett concept, the relationships between "the three great problems" become deeper.

world was reversible, while consciousness creates the sensing of time flow and the distinction between the present, past, and future (upper arrow in the right-hand part of Fig. 6.8). The present is distinguished by the fact that at this moment the subject is choosing the alternative that will be in the nearest future perceived by his consciousness. In the quantum world of inanimate matter, which evolves according to the Schrödinger equation, the notions of 'present', 'past', and 'future' are simply absent.

(3) As one of the aspects of the picture appearing in consciousness, the time flow singles out the time arrow. With respect to this time arrow, entropy increases. However, entropy decreases in the sphere occupied by life (living matter develops and becomes self-organized). This is also explained in the framework of the EEC.

Briefly, this is because consciousness (in the boundary state) perceives various alternatives, analyzes them, and modifies their subjective probabilities, preferring the alternatives that are more favorable for life. The last of these means that a subject perceiving some alternative is more probable to perceive, in the following instant of time, one of the alternatives that are most favorable for him. The special 'choice' of alternatives providing survival means that *the dynamics of life observed by consciousness are determined not by the cause but by the goal.* And this, of course, means a decrease in entropy in the life sphere.

(4) Identification of consciousness with the separation of alternatives actually generates a new *notion of quantum consciousness*, which has a unique property. Consciousness understood this way enters, as a necessary element, both quantum physics and psychology. This way, direct contact between these two sciences is established. Continuing this analysis, we see that quantum consciousness forms *the bridge between the natural sciences and the sphere of the humanities* (including nonscientific forms like

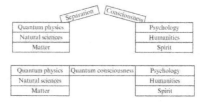

Fig. 6.9   Identification of consciousness with the separation of alternatives generates the new notion of quantum consciousness, which is a general subject of study and, hence, the bridge between the natural sciences and humanities, between matter and spirit.

religion). Eventually, one can say that quantum consciousness builds the bridge between matter and spirit (Fig. 6.9).

This is indeed a bridge over a chasm. There are many important relations between the material and spiritual spheres. However, quantum consciousness apparently makes a more solid contact between them: each of these spheres needs the other one for the sake of being conceptually closed.

6.9   Conclusion

The following conclusions can be drawn from above analysis.

(1) The question of *what happens during measurement* should be answered in the framework of physics as follows. In any case, reduction does not occur; what proceeds is the entanglement of the measured system with its environment and, as a consequence, decoherence of the measured system. This is derived strictly in the framework of quantum mechanics. From a somewhat broader viewpoint, the answer to the same question is: during measurement, the observer's consciousness perceives the measurement result, which is equivalent to separating the alternatives.

(2) The question of *what life is from the viewpoint of physics* should be answered in the following way. Since consciousness is identified with the separation of alternatives, 'quantum consciousness', namely, the concept of consciousness resulting from this analysis, erects *a bridge between physics and life*. Life cannot be explained by only physical processes which obey the laws of physics. At the same time, one cannot say that no relation exists between the phenomenon of life and the laws of physics. This relation exists and it is important, but it is not a direct relation. From the Extended Everett Concept, it follows that the 'quantum consciousness' throws a bridge between physics and life: *Quantum physics cannot do without the notion as*

consciousness (the most important component of the phenomenon of life), while *life cannot be explained without invoking quantum physics.*

(3) And, finally, the last question is: *where does irreversibility come from.* Based on the Extended Everett Concept, we come to the conclusion that *the objective (quantum) world is reversible, while (quantum) irreversibility arises in the picture of this world created by consciousness.* Consciousness builds its life in a picture of the world where the time arrow exists, there is a qualitative difference between past, present, and future, and the future is 'locally predictable'. This is, of course, not accidental, since a survival strategy, i.e. the very existence of life, is only possible in such a world. This possibility is realized by increasing the probability of the subject observing favorable alternatives and means an entropy reduction in the sphere of life.

PART 3

Parallel Scenarios and Sphere of Life

It has been shown in Part 2 that mystic features of consciousness (such as super-intuition and probabilistic miracles) follow from the fact that our world is quantum and can therefore be presented as the set of parallel classical worlds (called Everett's worlds). Consciousness can perform (probabilistic) miracles choosing the worlds favorable for living in them. This is the very essence of life.

The advantageous worlds are chosen when consciousness looks at the future of all alternative (classical) worlds forming the quantum world. In the present part we shall present this process in terms of *alternative scenarios* rather than the alternative worlds.

This allows one to efficiently describe evolution of life that is determined not only by causes but also by goals. This evolution law does not contradict the conventional causal laws of natural science because it acts only in the *sphere of life* defined as the set of scenarios favorable for life. The *principle of life* governing the evolution of living beings is formulated. It naturally leads to appearance of such phenomena and notions as providence, karma and God (or closely resmbling them). The present approach thus unifies natural sciences with the sphere of spiritual knowledge including religion.

The readers that are not professional physicists may skip Chapter 7.

Chapter 7

Evolution of life: goal instead of cause (for physicists)

Analysis performed in the preceding chapters on the basis of the Extended Everett's Concept led us to theory of consciousness including the ability of consciousness to choose the best of all possible scenarios of being. This is in fact the most essential feature of life as a special phenomenon. With this ability accounting, the evolution of life should be determined bot only by the initial conditions but also by the goals (aims) specific for life, first of all the aim of surviving.

The natural questions in this connection are, first, how can one formulate this new law of evolution for living beings and the world around them and. second, whether this new evolution law is compatible with the conventional laws of natural sciences. This chapter gives the answers to these questions along the lines of consideration outlined in the paper [Mensky (2007c)].

It is shown that EEC may be reformulated in terms of the mathematical operation of postcorrection. This operation is defined as correcting the state which guarantees certain characteristics in future. The criteria which may be used for postcorrection as well as the corresponding phenomena in the sphere of life are classified.

We tried to do the material of this chapter independent of the previous chapters, on the cost of brief repeating the main ideas. Sects. 7.2, 7.3 contain some mathematical formalism and may be recommended for the readers familiar with quantum mechanics. For the rest of the readers we formulate the main ideas in other sections without much mathematics.

7.1   Introduction

7.1.1   *Main ideas of Extended Everett's Concept*

From the time of creation of quantum mechanics up to nowadays conceptual problems of this theory, or paradoxes, are not solved. They are often formulated as *the problem of measurement.* Attempts of solving these problems are performed in the framework of various *interpretations of quantum mechanics.*

Commonly accepted is *Copenhagen interpretation* including the *reduction postulate* declaring that the state of a quantum system changes after its measurement to one of the alternative states corresponding to one of the alternative measurement outputs (readouts).

However, this assumption contradicts linearity of evolution in quantum mechanics. Indeed, the measuring procedure may be considered as an interaction between the measured system and another quantum system, called measuring device. It follows from linearity of quantum-mechanical evolution that the state of the system and the measuring device after the measurement (interaction) is a superposition including *all alternative states* (alternative measurement outputs) as the superposition components. Meanwhile the reduction postulate requires that only one of these alternative states should maintain after the measurement (interaction) is over.

In the interpretation of quantum mechanics suggested in 1957 by Hugh Everett [Everett (1957); DeWitt and Graham (1973)] linearity is taken as a basic principle. Therefore, *all alternatives coexist as the components of a superposition and therefore equally real* in the framework of this interpretation.

An observer always watches only a single alternative, that seemingly contradicts to the all alternatives being present in the state of the system and the measuring device. In the Everett's interpretation this is explained, or simply described, by existence of "many classical worlds" (corresponding to various alternatives) or, equivalently, by the formulation that *the observer's consciousness separates the alternatives* from each other (subjectively the observer, when watching some alternative, cannot watch the others).

In the Extended Everett's Concept (EEC) which has been proposed by the author [Mensky (2000a, 2005a,b, 2007b)], *the observer's consciousness is identified with separating alternatives.* This simplifies the logical structure of the theory and simultaneously leads to new consequences. Indeed, turning off the consciousness, i.e., going over to unconscious (in sleep, trance of meditation) means turning off the separation, i.e., the ability to watch all alternatives simultaneously. Therefore, in states like trance the explicit consciousness (plain consciousness) is lost, however the *super-consciousness* arises, i.e., the ability to take information from all alternatives, compare them with each other and choose the favorable one.[1]

This allows one to explain the well known phenomena of *free will*, *necessity of sleep* (for keeping health and life), as well as such unusual (but evidently existing) phenomena as *direct watching the truth* (which cannot be deduced from the reality provided by the explicit consciousness), among them *scientific insights*, and even *"control of reality"* in the form of "probabilistic miracles" (perceiving the events which are in principle possible but with very small probabilities).

Thus, according to EEC the material world is described by conventional linear quantum mechanics, but in the observer's consciousness this world looks is divided in various classical alternatives (alternative classical realities). In the state of explicit consciousness these alternatives are separated (i.e., they are watched by the explicit consciousness separately from each other).[2] However, when the explicit consciousness is turned off or weakened (in the regime of unconscious) all the alternatives are watched simultaneously, compared with each other, and the most favorable of them are chosen (probabilities of watching the rest of them becoming very low). As a result, the picture perceived subjectively looks as the control of reality, providing support of life.

7.1.2 *Scenarios favorable for life*

The principal feature of humans, and in fact of any living being, is, according to EEC, its ability, overcoming the separation of the alternatives (i.e., considering all the alternatives), to follow each of them up to the distant time moment in future, find out what alternatives provide survival and choose them for subjective perceiving, excluding the rest (making the probabilities of the rest very low).

It is clear that this ability to choose an favorable reality is a basic condition, or even simply definition, of life as such. In this context it is

---

[1]Let us make terminology more precise. With (explicit) consciousness turned off, the process of taking information from all classical realities (that can be called *super-cognition*) becomes possible. Taking the (part of) this *super-information* to the explicitly conscious state is what is called *super-consciousness*.

[2]In EEC separating alternatives is in fact a definition of the explicit consciousness.

inadequate to talk about consciousness. Instead, something more general may be meant that is connected not only with humans, but with all living beings, and even with life as a whole.

The essential part of this concept is that the *evolution of living matter (and its environment) is determined not only by causes, but also by the goals*, first of all by the goals of survival and improvement of the quality of life.

In the present chapter we shall introduce the mathematical formalism that describes this principal feature of living matter.

In the framework of Extended Everett's Concept this feature is presented as the ability (of human) to make more probable that scenario of the world (actually of the environment) which is estimated as favorable on the basis of the comparison (made in the state of unconscious) of all possible scenarios. Now, in a more wide context, we are concentrated on the final conclusion that the living beings can choose the favorable scenario for evolution of its body and the environment. This conclusion is referred to life as a whole and such specific feature as consciousness is already not assumed.[3]

In accord with this reasoning, we shall tell in this chapter about evolution of living matter, without mentioning consciousness and the Extended Everett's Concept. It will be assumed that the evolution of living matter includes correction providing survival at distant time moments. This correction leaves only favorable scenarios of evolution in the sphere of life. Unfavorable scenarios do not disappear from the (quantum) reality but are left outside the sphere of life.

From the mathematical viewpoint this correction (selection of favorable scenarios) is presented by the special operation which is called *postcorrection*. This operation (defined in the next section) depends on the *criteria of life quality* that may be chosen in different way. Various criteria for postcorrection leading to various aspects of the phenomenon of life are discussed in the subsequent sections.

---

[3]Let us recall that even in case of human the most important processes were realized in unconscious state, i.e., outside the sphere of consciousness. In fact, properties of the human consciousness were important for the given conclusion only as a transparent hint leading to the more general issue valid in a wider sphere.

7.2 Life as the postcorrection in the criterion of survival

Life is a phenomenon which is realized by living matter consisting of living organisms (living beings). Living matter differs from non-living matter in that its dynamics is determined not only by *cause* (i.e., the initial state), but also by *goal* (target, aim) i.e., by the state this matter would have in future. By the goal it is meant first of all *survival* (persistence of life). However, in case of sufficiently perfect forms of life more complicated goals are essential. They can be formulated in terms of *quality of life*.

Important features of the phenomenon of life are connected with consideration of the whole living matter and the balance between various living organisms. However, essential features of life may be illustrated in the course of consideration of a single living being and a collective of living beings. Let us analyze this simple situation.

7.2.1 *Notion of postcorrection*

Since a living being consists of atoms which interact with each other, it may be considered as a physical system. According to the modern views this is a quantum system. Let us apply the term *"living system"* to refer this quantum system. Let $\mathcal{H}$ is a space of states of this system. The state of the environment will be considered to be fixed.[4]

Let $\{L, D\}$ (from the initial letters of the words 'life' 'death') is a complete system of orthogonal projectors in the state space $\mathcal{H}$, so that $L + D = 1$ and $LD = 0$. These projectors determine two orthogonal and complimentary subspaces $L\mathcal{H}$ and $D\mathcal{H}$ in the whole space $\mathcal{H}$. The subspace $L\mathcal{H}$ is interpreted as the space of the states in which the body of the living being is acting properly (stays alive). The subspace $D\mathcal{H}$, vice versa, is interpreted as the space of the states in which the processes of life are seriosly violated, the living being is dead. The projector L will be called the *operator of survival*.

If the quantum system is in the state $|\psi(t_0)\rangle$ at a time moment t_0, then its state $|\psi\rangle = U(t, t_0)|\psi(t_0)\rangle$ at time t is determined by action of the unitary evolution operator $U(t, t_0) = U_{t-t_0}$. This description of evolution is characteristic of non-living matter, whose dynamics is determined by causes (in the given case, the initial state, if the Hamiltonianis assumed fixed). However, it is not enough for living matter. *The dynamics of living*

---

[4]This is a sufficiently good approximation if the changes caused by the influence of the living being on its environment is not essential for its life.

matter is partially determined by goals, i.e., by characteristics of the future state of the living being.

In the simplest case the goal is survival. This means that the state of our living system in a distant future should be in the subspace $L\mathcal{H}$. This is provided by periodical correction (selection) of initial conditions in such a way that the desired result be provided in future. Such correction (selection of favorable scenarios) may be called *postcorrection. The operation of postcorrection is a correction of the present state of the living system, but it is performed according to the criterion which is applied to the future state of the system.*

Remark. Usage of a "future state" for characterizing earlier states has been discussed, under name of *postselection*, by Y. Aharonov, P.G. Bergmann and J.L. Lebowitz in the paper published in 1964 [Aharonov, Bergmann, and Lebowitz (1964)] and by Y. Aharonov with coauthors in the subsequent works [Aharonov and Vaidman (1991); Aharonov and Gruss (2005)]. In the concept of postselection both an initial time and some later moment of time ("final time") are fixed. In [Aharonov, Bergmann, and Lebowitz (1964)] the formula for the probabilities of various outputs of the measurement performed at an intermediate time (between initial and final times) was derived. The operation of postcorrection differs from postselection in the mathematical viewpoint since 1) not a single state but a subspace (of arbitrary dimension) is fixed in future (at the "final time"), and 2) the initial state undergoes a correction. But the principal difference is in the physical interpretation of these concepts. In the most of the works devoted to postselection, or two-vector formalism, this concept was used for analyzing some events predicted by conventional quantum mechanics for usual quantum material systems. In the paper [Aharonov and Gruss (2005)]) a certain interpretation of quantum mechanics based upon postselection (two-vector formalism) was proposed, in which the choice of an output of a measurement was associated with the choice of a state vector in future. Instead of this, *postcorrection* describes not a usual material system, but a "living system" and its evolution.

7.2.2 *Simplest example of postcorrection*

Let us consider the simplest example of postcorrection. For simplicity of notation, we shall fix two time moments, "the present time" $t = t_0$ and "the future time" $t = t_0 + T$. Denote by U_T the evolution operator leading from the present time to the future time.

Let the living system's state at time $t = t_0$ be presented by the vector $|\psi\rangle \in \mathcal{H}$. If only usual (characteristic of non-living systems) dynamics acts, then after time interval τ the state vector should be $U_\tau|\psi\rangle$. However, life as a special phenomenon is described only by those scenarios in which the usual evolution provides survival (continuation of life). For life prolonging for the time T, it is necessary to restrict the initial conditions by the subspace $U_T^{-1}LU_T\mathcal{H}$.

Thus, the correction which selects favorable scenarios is described by the projector $L_T = U_T^{-1}LU_T$ which may be called *postcorrection operator*. The living system's evolution, with the postcorrection taken into account, may be presented as a series of short periods of the usual (causal) evolution, each of them preceded by postcorrection. This is described as the action of the operator

$$U_{\text{cor}} = U_\tau L_T \cdot \cdots \cdot U_\tau L_T \cdot U_\tau L_T. \qquad (7.1)$$

7.2.3   *Interpretation in terms of "life sphere"*

Selecting favorable scenarios does not mean violation of the laws of nature as such. *Material world is described as usual by all scenarios* that are obtained by the action of the unitary evolution operator $U(t, t_0)$ on the arbitrary initial state vectors. This is enough for depicting non-living matter. However, *the phenomenon of life is presented by only a part of all scenarios* of evolution. "Unfavorable" (for life) scenarios are left "outside the *sphere of life*".

A human being, as a living system (therefore restricted by the life sphere), has only favorable scenarios forming the picture appearing in his consciousness. Unconscious but living organism (not human being) is also in the sphere of life, and its reflection of the world is presented by favorable (for it) scenarios.[5] Subjectively (from the viewpoint of the human being or in the reflection of the simplest living being) this looks as if the living being could find out what should be its state in a distant time $t_0 + T$ and correct the state at time t_0 in such a way that provide prolongation of life at time $t_0 + T$.

---

[5]This expresses only the very principle of life, not complete description of it; accidents and other causes terminating life should also be included in the complete description of living matter.

7.2.4   *Postcorrection in terms of EEC*

It could be not quite clear in the above said that the words "unfavorable scenarios are left outside the sphere of life" mean. To make them more clear, let us reformulate the same in the language utilized in EEC, however with the help of the above mathematics.

In the framework of Extended Everett's Concept the (explicit) consciousness is identified with the separation of the alternatives. In the transition to the regime of unconscious ("at the edge of (explicit) consciousness") the separation of the alternatives disappears, and the possibility arises to compare all alternatives between each other, select favorable ones and discard the rest. How could this be expressed in the language of mathematical formulas?

Let the set of the (quasiclassical) alternatives at the present time be defined as the set of subspaces $\{\mathcal{H}_i\}$. Assume that the favorable (providing survival in the time interval T) are the alternatives $i \in I$, while the rest alternatives $i' \in \bar{I}$ (where $I \bigcup \bar{I}$ is the set of all alternatives) are unfavorable. This means that $LU_T\mathcal{H}_i = \mathcal{H}_i$ for $i \in I$ and $LU_T\mathcal{H}_{i'} = 0$ for $i' \in \bar{I}$. Then the postcorrection operator $L_T = U_T^{-1}LU_T$ is a projector on the subspace $\mathcal{H}_I$, which is the sum of "favorable" subspaces $\mathcal{H}_i$, $i \in I$.[6]

Therefore, "to stay in the sphere of life" means that only favorable (for life) alternatives are left in the picture appearing before the consciousness. The rest alternatives (subspaces) do not disappear (this would be the violation of the laws of nature), but simply are left outside the sphere embraced by the consciousness of human beings and the reflection of the world by the living being.

From this point of view the statement that the phenomenon of life is described by postcorrection in the criterion of survival is in fact not a postulate but only a *mathematical form of the definition of life*. Any reasonable definition should differ from it only in details, but not in principle. Indeed, the essence of the phenomenon of life is a strategy of survival, and the efficient survival is provided only by estimating the future of a living system (from the point of view of its survival) and corresponding correction of the system's present state.

---

[6]In Sect. 7.4 we shall see that the real situation is very close to this, differing only in that the sets I and $\bar{I}$ do not necessarily cover the set of all alternatives: the alternatives (subspaces) which are intermediate between completely favorable and completely unfavorable ones may exist.

7.2.5   *Other issues to be accounted*

Some remarks should be made about the evolution law (7.1).

- In the above specified formulas we assumed that the operator of causal evolution depends only on the time interval, but does not depend of the initial time moment: $U(t, t') = U_{t-t'}$. If the environment of the living being varies with time, this assumption is invalid and one has to make use of the evolution operator $U(t, t')$ depending on two arguments.

- We assumed that the evolution of the environment is given independently of the state of the living system. This may be justified in many cases. However, this assumption has to be abandoned in case of more sophisticated criteria for postcorrection (than the criterion of survival) which mayinclude parameters of the environment as well as the parameters of the living system itself (such criteria will be considered later on). Then $\mathcal{H}$ has to be defined as the space of states of the compound system including the living system and its environment. The operator $U(t, t')$ is then the evolution operator in this more wide space.

- The evolution presented by the operator (7.1) consists of the series of operations of the causal evolution and postcorrection. It is characterized by two time parameters: the *period of correction* τ and the *depth of postcorrection* T. It is possible that some processes in living organisms are adequately presented by such type of evolution (for example, higher animals and humans periodically experience the state of sleep in which the correction of the state of the organism is performed). However, continuous regime is typical for other correcting processes. In these cases an evolution law with continuous correction should be applied. The simplest variant of it can be obtained as a limit of the discrete process.

- We considered a transparent mathematical model of life in which the postcorrection is presented by a projector. This may be generalized. For example the postcorrection operator may be taken to be a positive operator (more general than a projector). This is evidently necessary for those criteria for postcorrection that are connected not with survival, but with less critical parameters of quality of life. Such criteria will be considered in Sect. 7.4.

Up to now we considered only the simplest scheme for support of life of a single living being. This scheme requires only a single criterion of life called survival and mathematically presented by the projector L. This may be enough for primitive forms of life in the condition of unlimiting resources (first of all food). However, for realistic description of more sophisticated forms of life one has to consider more complicated criteria. Besides, the role played by the living beings in respect to each other should be taken into account.

All this requires further generalizations of the mathematical model of life. Not pretending to be quite general and precise in detail, we shall illustrate possibilities of such generalizations in some typical situations. In Sect. 7.3 a sort of collective criterion of survival will be considered, and in Sect. 7.4 the classification of various criteria of life and corresponding aspects of the phenomenon of life will be presented.

7.3   Collective strategy of survival

The model duscussed in Sect. 7.2 shows how the principal goal of a single living being, survival, may be described mathematically in terms of the special operation, postcorrection, with the help of the criterion of survival of this living being. This criterion was presented by the projector L onto the subspace of those states of the living system that are interpreted as the states in which it remains alive. This model may be good in case of simple forms of life and unlimited resources (first of all the amount of food). Let us consider now *the model of life in which resources are limited so that only limited number of living being can survive.*

It is clear that in this case the relations between various living beings become important and should be taken into account. One possible strategy for survival of living beings in these difficult conditions is fighting them with each other. However, the *collective strategy of survival* is also possible in this case. Let us consider the simplest mathematical model of such a collective strategy.

Consider a collective consisting of N similar living beings (living systems), enumerated by the index $i \in \Omega$, where $\Omega = \{1, 2, \ldots, N\}$. The living system having number i is described by the operator of survival L_i (the corresponding orthogonal projectot being D_i). To formulate the model, introduce also the notations $L_I = \prod_{i \in I} L_i$ and $D_I = \prod_{i \in I} D_i$. Denote by $|I|$ the number of elements in the set I and by $\bar{I} = \Omega \setminus I$ the set of elements in Ω that are not elements of I.

In the conditions of unlimiting resources all living systems forming the collective can exist (survive) independently of each other. Then each of them may be described with the simple model considered in Sect. 7.2 so that all of them can survive forever. Assume however that the resources (for example food) that can be found in the environment are limited and their amount is sufficient only for survival of n living systems. In this case such control of the life of the collective may exist that takes into account the interests of the whole collective. Then a sort of a *"superorganism"* exists. This means that the collective consisting of N living beings behaves as a single living system. What has to be taken as an operator of survival of the whole collective in this case?

The simplest form of the *collective operator of survival* is

$$L_{(n)} = \sum_{I \subset \Omega, \, |I| = n} L_I D_{\bar{I}}$$

It is not difficult to show that this operator is a projector, and the projectors $L_{(n)}$ $L_{(n')}$ are orthogonal for $n \neq n'$. The set of projectors $\{L_{(n)} | n = 0, 1, 2, \ldots, N\}$ forms a complete system of orthogonal projectors.

The correction described by the operator of survival $L_{(n)}$ guarantees that in the time interval T precisely n living systems will be alive, the rest will be dead. This means that the resources will be sufficient for those which are alive. *The death of some members of the collective is in this case a condition for survival of the rest.*

It is interesting in such a model that the correction of the collective of the living systems expressed by the collective operator of survival $L_{(n)}$ describes not fighting of the members of the collective between each other, but the collective regulation providing survival of such number of organisms that the resources (food) is enough for them. It is not fighting because the state of each living system is corrected at the present time moment, and the corrected in this way states, simply because of the natural evolution (described by the operator U_T) lead to death of just that number of the members of the collective which is necessary for the rest being alive. Such correction of the state may be called *collective program of death* some members of the collective *for life* of the rest. The collective program of death does not determine which concrete members of the collective have to die, only their number. Therefore, this is actually the *strategy of collective survival* discriminating none.

Evidently, in some (if not all) collectives of animals the survival is regulated by collective criteria. This explains particularly the absence of intraspecific competition. In this respect the human collectives radically differ. Almost in any human group the collective criteria of various levels exist: for a nation, for a social group, for a family and so all, up to the individual criteria for single people. This leads to fighting or conflicts between different groups of people and in the limit to fighting all against all.

We shall see in Sect. 9.1 how existence of the universal (common for all) collective criterion may prevent the global crisis which otherwise could become inevitable perspective of the people on Earth.

7.4   Various criteria for postcorrection

In the preceding sections we considered the operation of postcorrection according to the *criterion of survival*, the most important criterion for living beings. In fact this criterion defines life as such. Evidently, the postcorrection of the simplest forms of life occurs according to only this criterion. However, for more sophisticated forms of life other criteria, which characterize the *quality of life* in more detail, become actual. Postcorrection may occur according to several various criteria acting together. Probably, the postcorrection of the human beings may be performed according to the criteria connected not only with their bodies but also with the parameters of the environment.

Investigation of various *postcorrection criteria* is an interesting problem that may be considered from various viewpoints. Not pretending for generality and validity of details, we can propose the following rough classification of the possible criteria.

- Criteria of survival

    - Criterion of survival for a single creature
    - Criterion of survival for a collective of creatures
    - Criterion of survival for the living matter as a whole

- Parameters of the state of the body

    - Evidence of being alive or dead
    - Various levels of the quality of life
    - Immaterial parameters (insignificant for the quality of life)

- Parameters of the environment (life conditions)

    - Parameters that are essential for surviving
    - Parameters that are essential for the quality of life
    - Immaterial parameters (insignificant for the quality of life)

Let us make some remarks concerning this (of course, oversimplified and approximate) scheme of classification.

It is clear that the sophistication of the living systems allows them to control not only their survival, but also proper quality of life. This may be achieved according to the same scheme of postcorrection as the scheme providing survival, but with projecting on a more narrow space of states in which not only life carries on but the quality of life stays sufficiently high. Formally this means that in any state from the given subspace some *parameters of the state of the body* characterizing the quality of life are in the given limits.

The question naturally arises why we included *immaterial parameters* in the list of the postcorrection criteria. Indeed, the control on them is not necessary from the viewpoint of the internal needs of life. However, the experience points out that at least human beings realize such a control that reveals itself in the phenomenon of *free will*. Indeed, a person can to choose one or another variant of behavior according to his will, not changing by this choice essentially neither the fact of survival, nor even the quality of life. For example, he may in certain limits choose time of meals and amount of food. The more so, one may to decide quite arbitrarily whether he opens or closes the window, read a book or watch TV and so on. In the framework of our model the free will, or the choice of immaterial parameters of the body may be described as the postcorrection for a short time interval.

The postcorrection performed according to the parameter of the environment is just what has been called in the preceding chapters probabilistic miracles. This type of correction leads to the realization of such events that can otherwise happen only with low probabilities. If the corresponding parameters are important for survival, the postcorrection may provide what looks as providential escape. If the parameters concern life quality, the result of the postcorrection may be a bit of luck. The postcorrection with immaterial parameters may result in an arbitrary control on the environment that is most closely resembles the miracles of fairy tails.

Let us remark that happenings looking as miracles do not, strictly speaking, contradict natural laws. Indeed, in quantum mechanics always (and often also in classical physics) results of observations may be predicted only as random events. Even if some event is to happen with low probability, its actual occurring does not contradict the probabilistic law. In case of such occurring one may interpret it as a rare coincidence or (if the event had been expected as fulfilling someone's will) a probabilistic miracle.

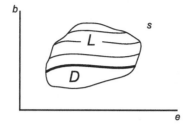

Fig. 7.1 Various criteria for postcorrection: state of the word s is determined by the state of the environment e and the state of the body b. The regions L and D correspond to survival and death. Horizontal lines separate the regions corresponding to different levels of the quality of life. Any subregion on the plane determines a criterion according to which the postcorrection may be performed (but is not necessarily performed).

Probabilistic miracle, i.e. arbitrary realization of events which have low probability although possible in principle does not seem necessary for life in the usual meaning of the word. However, first, this phenomenon naturally enters the general scheme so that its exclusion could look artificial, and, secondly, the human experience seems to point out that the events of this type really take place (see Chapter 2).

Considering various parameters for postcorrection from somewhat different point of view, one may suggest the following (of course, also quite tentative) classification (see Fig. 7.1). Denote by s (after the word "states") the set of various parameters of life (characterizing both the body and the environment). The parameter s is in fact a pair $s = (e, b)$, where e (after the word "environment") characterize life conditions (i.e., the state of the environment) and corresponds to the horizontal axis, while the parameter b (after the word "body") refers to the state of the body of the living being (or the bodies of the collective of the living beings) and corresponds to the vertical axis. The parameter s lies in some two-dimensional region, in which one may talk about life. This region is separated by a horizontal line in two subregions. The upper part corresponds survival L, and the lower part to death D. The region of survival in turn is separated in the subregions corresponding to various levels of the quality of life.[7]

Each subregion in the upper part of the region drawn in Fig. 7.1) determines some criterion according to which the postcorrection may in principle be performed (but is not necessarily performed). Of course, in some cases the formulation of the postcorrection requires the space of states of the

---

[7]In reality each of the parameters e and b is multidimensional, thus one may talk of the two-dimensional region only for more lucidity.

whole world, but not only the living system itself as in the examples discussed above. This means that one has to consider the evolution of the system including both living system and its environment.

The postcorrection according to various criteria describes various aspects of the phenomenon of life which can be characterized in the following way.

- *Life* = postcorrection in the criterion of survival for the living matter as a whole.
- *Survival* = postcorrection in the criterion of survival relating to the body (bodies).
- *Support of health* = postcorrection in the criterion of quality of life relating to the body.
- *Free will* = postcorrection in the criterion, relating to the own body, but as a rule immaterial for survival.[8]
- *Control on the appearing reality (probabilistic miracle)* = postcorrection in the criterion relating to an object outside the own body.

7.4.1 *Postcorrection providing super-intuition*

There is one more class of unusual phenomena in the sphere of consciousness (and therefore in the sphere of life) that can be explained by postcorrection:

- *Insight* = postcorrection according to the criterion of truth

This class includes foresights, insights (among them scientific insights). They may be generally characterized as *super-intuition* or *direct vision of truth*, i.e., conclusions not supported by logic and/or facts. These phenomena can be explained by postcorrection.

Explain first how the phenomenon under consideration happens. Let someone formulate a question or poses a problem (a scientific problem is a good example). The question or the problem turns out so difficult that it cannot be answered (solved) with the help of the usual rational consideration based upon the known data and knowledge. Then one more way exists for solving the problem. One may go over to the regime of unconscious (not necessarily completely turning off explicit consciousness, but disconnecting it from the given problem). In this regime the genuine solution of the problem comes unexpectedly and without any further efforts, as an insight.

---

[8]The exclusions such as a suicide require more detailed model accounting for the action of the living system onto its environment.

Not each one happened to observe this sequence of events. Some people might actually meet with them but not properly interpret because it is natural to interpret the unexpected guess as the last step of the previous logical consideration of the problem. However, many great scientists evidenced that some of their most interesting insights could not be based upon the previously known facts and the conventional procedures of analysis. Completely new ideas appeared in these cases "from nowhere", in non-verbal form, and were accompanied by complete confidence in their validity. The same felt sometimes people of other professions according to the questions and problems of various character.

How such things may happen? We shall show that the explanation may be based on the phenomenon of postcorrection. For simplicity suppose that the problem reduces to the choice between several already formulated "solutions", most of which are wrong. If there is no way to choose right solution with the help of the logical procedure and the known facts, then no way to solve the problem correctly seems to exist. How then postcorrection may help to find out what of the presupposed variants of the solution is right?

The right selection may be performed in this case according to some genuine criterion of truth (not yet known) with the help of postcorrection. Even if the problem cannot be solved at the present time by conventional methods (on the basis of the known facts and logical conclusions), it may have evident solution in future. For example, some future events may point to the correct solution. In case of scientific problem some new experiments may be realized in future that, from all seemingly possible solutions of the problem unambiguously point to the single right solution.

In all such cases the postcorrection results in the choice of the genuine solution of the problem, which will be confirmed in the future. The man, when being in the regime of unconscious, obtains the ability to look into the future, and makes use of the obtained information. Returning to conscious state, he obtains finally the right solution in the present.

The idea may be clarified if it is reformulated in terms of the states of brain. Cpnsider the states of the brain corresponding to various selections of "probe solutions" of the problem (wrong ideas of the solution among them). Let the postcorrection is applied to the superposition of these states. The postcorrection is performed according to the criterion of right solution of the problem, the criterion which will exist in future.

Therefore, the operator of postcorrection projects the state of the brain (presented as the superposition) onto the single component of the superposition, and just that component which will turn out in the future to correspond to the criterion of correct solution. Therefore, with the post-correction taken into account, the brain is in the state of right solution already in the present moment. The man feels as if he knows which solution is right although he does not understand where this feeling came from. Insight, or direct vision of truth, arises as a result of postcorrection. The man is in this case absolutely sure that the selection made by him in the course of the insight is truth.

Great scientists, Albert Einstein among them, confirm the fact that they always feel absolute confidence in the solution found in the insight, and the solution found in this way always turns out correct in the course of its verification by conventional methods. By the way, the "formal proof" that the scientist always performs after instantaneous insight, may be just that criterion of truth which does not exists (not yet found) at the moment of the insight but arises later on.

Solving a problem is easier if it is known that the solution exists and much easier if the final formula (not its proof) is available. Just this situation of the solution (not proof) known beforehand realizes in the scientific insight and subsequent derivation of the formal proof for the foreseen solution. Indeed, the scientist anticipates the right solution in the course of insight, and it becomes much easier for him to find a formal solution of the problem. It is curious that in this case the scientist foresees the certainly right answer which himself will find in some time.[9]

The operator of postcorrection selecting the right solution of the problem may be presented in the form $P_T = U_T^{-1} P U_T$, where P is a criterion of the correct solution. The operation of postcorrection presented by the operator P_T is efficient if the criterion P is not realizable at present, but can be realized in the time interval T.

This leads us to the question about the *role of brain*. Many attempts to explain how work of brain can produce the phenomenon of consciousness gave in fact no result. In each of these attempts either a logical circle is included (what should be proved is implicitly assumed) or not the phenomenon of consciousness as such is defined but various operations performed in the conscious state (for example, calculations or logical conclusions).

From the point of view of the concept under consideration, consciousness is not a product of brain, but a separate, independent phenomenon, closely

---

[9]This ability is very exciting in case of great scientists, but it is often is exploited by many experienced scientists as well as people from other professions and simple people in the each-day life.

connected with the very concept of life. What about brain, it is an instrument used by consciousness to control the body and obtain information of its state (and through its organs, about the state of the environment). In the other words, brain (or rather some structures in it) is the part of the body that realizes its contact with consciousness. It is an interface between consciousness and the body as a whole. In particular, when it is necessary the brain forms the queries that should later be answered. Sometimes these queries are answered by the brain itself with the help of the processes of the type of calculations and logical operations. Other queries cannot be solved directly in brain and are solved with the help of "direct vision of truth" (e.g. by postcorrection).

Remark 7.1. A. Lossev and I. Novikov noted [Lossev and Novikov (1992)] that *time machines* (space-times including closed timelike curves), in case if they exist, might be used for solving mathematical problems with the help of the methods or technical devices which are not known at present but can be realized in future. For this aim, the problem is solved at the time when the necessary methods are created and then its solution is sent into the past with the help of the time machine. The above formulated mechanism for solving problems (of arbitrary types) with the help of postcorrection is quite analogous. The only difference is that the "time machine" acting in this process is virtual and "exists" only in human consciousness.

7.5　Conclusion

Although Extended Everett's Concept (EEC) originated in the attempt to improve the interpretation of quantum mechanics proposed by Everett, it is not a simply novel interpretation but in fact a theory going out of the framework of quantum mechanics. In EEC the assumption is accepted that the conventional (causal) laws of nature are insufficient for describing the phenomenon of life. The laws of nature elaborated by physics (including quantum physics), chemistry and other natural sciences govern behavior of non-living matter. It is impossible to explain the behavior of "living matter" on the basis of only conventional laws of nature.

It turns out however that the attentive analysis of quantum mechanics points out at least the principal points in which the laws acting in the sphere of life have to differ from physical laws, and allows to formulate these laws in their most general aspects. This analysis may be based on the Everett's many-worlds interpretation of quantum mechanics as well as on Extended Everett's Concept.

The evolution in the sphere of life depends not only on the causes (initial conditions) but also on the *goals of life*. The main goal existing always in connection with living beings is the goal of survival, or continuation of life (this may be survival of a single living being, or of some collective, for example of a herd or specie of animals). Less universal goal (but typical for humans) is improving quality of life, in particularly support of good health.

This type of evolution, accounting the goal (i.e., the state that should be achieved in future) can be presented with the help of the mathematical operation called *postcorrection*.

This operation corrects the state of a "living system" providing in future survival or even high quality of life (for example health). Introduction of such a formal description of the living system's evolution allows to classify various forms of life and various aspects of the phenomenon of life depending on what characteristics of life can be provided by such a correction. We shortly discussed only the key points of this classification.

The operation of postcorrection simplifies the logical structure of the theory following from EEC. In fact, it is sufficient to postulate that *the boundaries of the sphere of life are governed by postcorrection*. After this, the specification of the theory requires only the choice of criteria, according to which the postcorrection is performed.

The operation of postcorrection is connected, from the mathematical viewpoint, with the concept of postselection introduced in 1964 by Aharonov et al. [Aharonov, Bergmann, and Lebowitz (1964)]. However, in the framework of our approach, this operation obtains a quite unexpected (for physics) interpretation, as describing evolution of living matter. This interpretation became possible because we did not restrict ourselves by the framework of physical laws as they were formulated for inanimate matter. Starting from the arguments originated in physics (conceptual problems of quantum mechanics) and following the ideas of EEC, we went behind the limits of proper physics and formulated the principal contours of theory of living matter.

Instead of the known assertion of physics that each event has its own cause, one has to agree that all important events and processes in the *sphere of life* are determined not only by causes but also by goals, first of all by the goal of survival. In the resulting theory the operation of postcorrection is a mathematical formalization of the almost evident fact that the goals play central role in evolution of living matter.

Theory of consciousness and life sketched in the framework of EEC essentially differs from the usual mechanistic approach where the phenomenon of consciousness is considered as a function of brain. From the viewpoint of theory of "quantum consciousness" originating from EEC the brain is rather an instrument exploited by consciousness (as a specific feature of a "living system") for the control over the body and obtaining information about the environment through the body and its organs.

This, by the way, allows one to look at the problem of *artificial intellect* in another way. The conclusion following from this viewpoint is that it is possible to create an automate possessing intellectual abilities (there are great achievements in this respect nowadays), but it is principally impossible to create machine having consciousness as something that can perform postcorrection, i.e., such that can be called "artificial living being".

Postcorrection is a good illustration of the question of irreversibility considered in Chapter 6. Inanimate matter evolves according to Schrödinger equation that is symmetric in time, but the evolution of living matter includes the irreversible operation of postcorrection.

Let us make finally one more remark that demonstrates how natural is the law of evolution of the living system (7.1) including postcorrection. This law is, in its spirit, very similar to what is called the *anthropic principle*. The anthropic principle explains the special "fine tuning" of the parameters of our world by the fact that in case of any other set of the parameters organic life would be impossible and therefore impossible were the situation where human beings could observe this world. The principle of life, formulated as ability of postcorrection in the criterion of survival, means in fact something quite similar, even in the softer variant.

In order to explain this, we have to underline once more that *the postcorrection describes selecting those scenarios which keep to stay in the sphere of life*. The rest scenarios do not disappear. They are just as real as those selected, but they are not included in the sphere of life, i.e. an observer cannot watch these "unfavorable for life" scenarios. *"The sphere of life" is the image of our world which an observer may see.* If just this image (i.e. not "the whole world" but only "the sphere of life") is taken as a starting point of the logical construction, then the result of the construction will necessarily be the selection of only favorable for life scenarios as the law of evolution of the living matter.[10]

---

[10]Let us remark that the life sphere is definded as the set of the scenarios that are favorable for life as a whole but not for the individual living beings. Vice versa, in definite situations the death of some living being may be the condition for surviving of great number of living beings, and then the scenario supposing this death is in the life sphere. Moreover, the death of each living being when growing old is evidently the necessary condition of surviving the corresponding species and therefore the life sphere

Thus, postcorrection in the evolution of the living matter (of the sphere of life) does not need even postulating. Instead of this it may be derived from the (generalized) anthropic principle. Non-living matter satisfies the usual quantum-mechanical evolution law. The evolution of the living matter (of the sphere of life) simply by definition should include postcorrection.

---

includes the scenarios supposing death of all living beings (may be excluding the simplest of them) with the increase of years.

Chapter 8

Life in terms of alternative scenarios instead of parallel worlds

The special features of consciousness (usually treated as mystical ones) are explain above as a consequence of the quantum character of reality that can be presented as coexisting of various classical realities, or parallel classical worlds. If we are interested in the general phenomenon of life (rather than the more narrow phenomenon of consciousness), it is more convenient to tell not only about the alternative classical worlds but also about alternative scenarios, i.e., various chains of the alternative worlds, one for each time moment.

In Chapter 7 the phenomenon of life in the set of all alternative scenarios has been described in terms of the mathematical operation of *postcorrection*. Here we shall present the same in verbal form, without special mathematical formalism.

The idea is that life is a special ability of living beings to choose those scenarios that are favorable for surviving and even for improving quality of life. This leads to the notion of *life sphere* as a subset of favorable scenarios. This notion allows one to express the qualitative difference of the laws of evolution acting in the inanimate matter from the laws of evolution of living matter.

8.1 Alternative worlds and alternative scenarios

The main idea of this book is that our world is quantum in its nature that may be presented as a set of coexisting (parallel) classical worlds, or classical realities. Consciousness separates these worlds so that subjectively the illusion appears of only a single world existing. In the next time moment also only one of the set of all alternative possible worlds appears in the subjective perception of the observer, and so on. Thus, with time passing, an alternative scenario (the chain of alternative worlds in the subsequent time moments) arises in the subjective perception of the observer. Objectively however all possible alternative scenarios coexist.

In order to understand the phenomenon of life, the set of alternative scenarios is more convenient than the set of alternative classical realities referring the definite time moment. Indeed, the scenarios which are favorable for life are naturally defined as those leading to surviving or, in a more general case, to an appropriate quality of life.

The law of life is choosing the favorable scenario (of course, in the framework of the definite restrictions). Therefore, only favorable scenarios should be included in the description of life. They form the subset of all scenarios that may be called *life sphere*. This is the cardinal difference from what we have in the description of inanimated matter where all scenarios act.

8.2   Evolution governed by goals

According to the conception of consciousness developed above, in the regime of unconscious a human, or generally a living being, have access to all parallel words, or all possible realities, can choose those realities that are favorable for surviving and for quality of life, and increase the subjective probabilities of the favorable worlds. Telling about scenarios (i.e., the chains of the alternative realities, one for each time moment), this means that only favorable scenarios are subjectively observed.

The resulting evolutions law takes into account not only the initial state, but also the future state of the "living system". According to such an evolution law, evolution depends not only on causes, but also on the goals (aims), first of all the goal to survive and have good quality of life.

In Chapter 7 this evolution law is expressed mathematically with the help of the operation of postcorrection. Here we shall try to restrict ourselves by the simple formulations, without mathematics.

Taking it in a simple form, postcorrection is excluding those initial states that leads to inappropriate states in future (for example, to death). Evolution including this operation is a usual evolution given by the conventional natural laws (say, quantum-mechanical Schrödinger's equation) but with such initial conditions that lead to the favorable final state.[1] As a result,

---

[1] Of course, the cases of inevitable deaths should be also accounting as well as the deaths that may be avoided but on the cost of the lower quality of life of the whole collective, see Chapter 7 for details.

evolution of life may be expressed as the set of favorable scenarios instead of the wider set of all possible (according to the natural laws) scenarios.

Thus, the evolution of inanimated matter is presented by the set of all possible scenarios, while the evolution of life is presented by the subset of the favorable scenarios. This subset of scenarios may be called *sphere of life*, of *life sphere*. In this formulation, the evolution law is already included in the definition of the corresponding set of scenarios, wider one for inanimated matter, narrower one for life.

8.3 "Principle of life"

The statement that only favorable scenarios (i.e., those forming the life sphere) are realized for living beings, may be called the *principle of life*. Yet one has to recall that this was derived from the properties of consciousness (as they are presented by the Extended Everett's Concept) and therefore refer to the sphere of subjective. The law thus formulated describes the evolution of life as it is perceived by the consciousness of human or reflected somehow in the process of functioning of living beings. This is the law of subjectively perceived evolution.

Thus, we may suggest 1) the simple but not quite precise or 2) the sophisticated but more precise formulations of the evolution laws:[2]

(1) Inanimated matter evolves according to all possible scenarios, while life evolves according to the scenarios belonging to the subset called life sphere.

(2) Objectively all matter evolves according to all possible scenarios, while subjective perception (reflection) of humans and generally living beings gives the picture of evolving both themselves and their environment according to the scenarios belonging to the life sphere.

8.4 Life principle as the generalization of the antropic principle

Let us make a remark demonstrating how natural for living systems this evolution law is. This law is, in its spirit, very similar to what is called *antropic principle* now widely applied in cosmology.

---

[2]see Chapter 7 for more details, however with some mathematics.

The parameters of our world (such as the masses of elementary particles and so on) prove to be "fine tuned". This means that these parameters lie in a very narrow interval which is necessary for existing the organic life. If the parameter of the world were not in this interval, life (at least of the kind we know) would be impossible. The question may be posed (and really posed by some authors) as to why it happened in this favorable way? Why the real world is just this?

The antropic principle explains "fine tuning" of the parameters of our (i.e., observed by us) world by the evident fact that in case of any other set of the parameters organic life would not be feasible and therefore no humans could exist to observe such a hypothetical world.

The principle of life, formulated above has just the same logical structure, it id quite similar to antropic principle [Mensky (2007c)]. The principle of life claims that life (together with the whole world as its environment) evolves according to the scenarios from the life sphere. This means that the evolution scenarios are favorable for life.[3] But this suggests in fact something quite similar to the antropic principle (even less imperative, because it refers only to subjectively perceived reality).

Let us underline once more that the principle of life is valid only for subjective aspect of reality. This feature of the principle shows itself very clearly in the application to the problem of global crises that we are considering below.

8.4.1 *Providence, karma, God*

In respect to life as a whole the above formulated principle of life looks as the *hand of God*: some higher force is concerned about the good destiny of all living beings, providing the evolution of themselves and the world, as they reflect it by their sense organs, the most favorable.

In the application to an individual person, the *principle of life* is parallel to the concept of *providence*, or divine disposal.

Remark however that the destiny of each person provided by the providence, or life principle, is not necessary favorable. The goal governing the evolution is care for life as a whole, for all living beings. The destiny of the give human should therefore be favorable only in case if his prosperity is favorable for people around him and even for life around him. This is the case if this person is kind to people, animals and life as a whole. We

---

[3]in the formulation of Chapter 7, living systems postcorrect their states to provide there survival or improve life quality.

may say about such person that he/she has altruistic (or even panhuman, ecological) consciousness. God cares for such people.

In other words, God provides a good destiny for good people, and in this sense "God is love". This is in accord with the concept of *karma* in Indian philosophy, that assumes the influence of an individual's past actions on his future lives.

8.4.2   *The answers of super-consciousness depend on the conscious life criteria*

Indications, obtained from "the other world" with the aid of the super-consciousness, give always precise answers to the presented question. However, the presented questions are formed in the conscious state and are therefore limited by the possibilities of the visible world. In particular, the questions are limited by the criteria of this person in his/her life.

As a result, precise answers to the presented questions do not guarantee, that the person, who is guided by these answers, comes to the solutions, which are truth in the general human sense. Of course, this does not guarantee the solutions which are correct in the sense of life as a whole.

A person, who is guided by such super-intuitive answers obtained from unconscious, does not compulsorily conduct righteous life. In order to conduct correct life, criteria themselves, by which man is guided, must be general (pan-human) and, moreover, ecological. Going over to such criteria, or criteria of life, is the work of the human on his consciousness. If this work is done and the consciousness changed, then answers of unconscious will ship the man to such solutions, which go for the good of entire living. And here then he/she will conduct life righteous, and here then the surrounding world will be always favorable to him/her (certainly, with exception of the cases, when for the comforts of the life as a whole is required victim from his/her side).

PART 4

Speculations or further development of the concept

In this part we will discuss some possibilities to develop the Quantum Concept of Life (QCL), presented in the previous parts, allowing more speculative considerations. One cannot speak about these considerations that they completely naturally follow from quantum physics. However, they are in accord with physics, and the conclusions, which appear as a result, make it possible that QCL is closer to the world religious confessions and well-known spiritual schools. In particular, we will attempt to interpret in the terms QCL such concepts as soul, paradise and hell.

Chapter 9

Escaping global crisis and life after death

If one allows himself to argue with more elements of fantasy, somewhat speculatively, then some concepts developed in the sphere of spiritual doctrines may be interpreted also in terms of the Quantum Concept of Consciousness and Quantum Concept of Life. We shall give here examples of the considerations of this type. While in the previous chapters we tried to make use of the ideas of quantum physics as the origin of the concept of consciousness and life, in the following consideration we shall sometimes appeal to the ideas of spiritual schools, first of all religious confessions as the origin pointing the direction to the further development of our conception.

9.1 Global crisis and eluding it (hell and paradise)

Science and its technological applications are usually considered to be the great achievement of mankind. However, the extremely rapid development of science in 21st century leads to global problems because new technologies are used by some people against other people. In the case of rapid technological progress, this is pregnant with a global catastrophe. The only way to prevent the cataclasm seems to be change of the egotistic consciousness of people, making it universal, panhuman. Theory of consciousness following from quantum mechanics seems to give us hope to do this.

Complete theory of consciousness becomes very important in the 21st century. This may be the most important scientific task in our time. The vital need of mankind seems to be in change of the consciousness. This requires the very deep understanding of the nature of consciousness, and adequate theory of this phenomenon.

9.1.1 *The global crisis: technical aspect*

Scientific achievements were always accompanied by problems, but in the years following each concrete step of scientific progress, these problems were always overcome by science itself (and could not be overcome otherwise). Thus arguments for admiring science arose twice: first at the moment of the achievement and later once more when its negative consequences having been overcome. The fact that the problems which were happily eliminated by scientific methods had been previously posed by the same science, were already forgotten at the happy time of the victory over these problems.

The most demonstrative example is nuclear weapon. Its creation meant that the life at Earth confronted the danger of its complete destruction, but the following development of enormous amount of the nuclear weapons led to the danger of guaranteed complete destruction of the two main nuclear-weapon states, USSR and USA, so that the resort to the nuclear weapon was prevented or at least seriously postponed. Of course there are many other examples, each one may recall some of them.

However this relatively admissible situation is gradually changing now when both scientific achievements and the problems resulting from these achievements emerge very often. Time for solution of a problem becomes insufficient because the next problem emerge. The number of the unsolved problems increases, opening a door for a global crisis.

The near-term perspective of a global crisis may be confirmed by scientific methods. There are scientific works showing that the radical changes in evolution of mankind that have been earlier extremely rare become with time more and more often. If the quantitative description of the series of these changes is extrapolated in future, the period between the radical changes should become practically null in the next few decades. [Panov (2008)] This means that the whole character of the evolution of mankind has to alter, that can be interpreted as prediction of a global crisis of some type. It is highly probable that this should be technological crisis.

Indeed, some of the key changes in the course of the evolution were connected with scientific achievements (scientific revolutions). Some of the recent scientific achievements (such as creation of personal computers or radical improvement of their efficiency) are of course examples of the key changes in the mankind evolution. Decreasing of the intervals between the scientific revolutions, accompanied by the corresponding social problems, is then a manifestation of the general law of the decreasing intervals between the key moments of the evolution.

In the light of these arguments the global crisis seems to be unavoidable. We shall argue that consciousness plays the key role at the moment of the crisis.

9.1.2   *Corrupted consciousness as an origin of the crisis*

Historical analysis shows that the scientific and technical development itself does not lead to negative consequences. The problems arise not as the direct consequence of the scientific discoveries and following technical achievements, but rather in connection with their applications not predicted at the moment of the discovery.

New technologies promise and in fact provide advantages for people if applied as the creators of them planned it. However some people guess that these technologies can be exploited against other people, for example as new types of weapons or as instruments of power over them. This implies new tensions in the society and sometimes serious dangers for it.

Therefore, the problems arise because of the corrupted consciousness of some people, those who care of only their own personal or group interests, who neglect interests of all others.[1]

9.1.3   *Change of consciousness for preventing the catastrophe*

In principle this points out clearly how the problems connected with science, and particularly the global crisis, may be prevented. Consciousness of human beings, egotistic in its nature now, has to be made altruistic. Such new consciousness may be called panhuman. Moreover, the new consciousness should be such that human beings care not only about human beings but also about all living beings. Such consciousness may be called ecological.

Such a change of consciousness is a vital necessity for our technical civilization. This has been clearly understood and unambiguously formulated by some thinkers in various periods of history. [Satprem (1970); Teilhard de Chardin (1959)]

Yet it was never clearly explained how this change of consciousness may be achieved. Complete theory of consciousness, taking into account all its features seems necessary for answering this challenge. We shall argue that

---

[1]Remark that here we, for the sake of simplicity, use the world "consciousness" in the wider sense than in other part of the book. However, this does not lead to conceptual mistake. The terminology may be made self-consistent if we explain the point in more details, but we avoid this for the sake of briefness.

the theory of consciousness following from quantum mechanics solves these questions in a very unexpected way and lead to optimistic conclusions. [Mensky (2007c)]

9.1.4   *Resolution of the crisis: paradise and hell at Earth*

Analysis of the situation by the usual methods shows that only a miracle may save mankind which came to the edge of the catastrophe. Analysis, based on "quantum theory of consciousness" shows that consciousness itself may create a sort of a miracle and that the catastrophe will be prevented with complete guarantee for those humans who will manage to make their consciousness altruistic.

Let us express this in a form of a bright metaphor. There are two alternative ways of the evolution for mankind as a whole: 1) the hell of the global crisis and the ruin of the whole mankind, and 2) the paradise of surviving mankind due to the global change of consciousness of all people into the panhuman one and thus preventing the catastrophe.[2] These two alternatives are clearly seen even with the existing theory. The very strange conclusions of the "quantum theory of consciousness" are following: i) both alternatives will objectively coexist, and ii) subjectively, all people who will manage to make their consciousness panhuman, will find themselves in the paradise of surviving mankind (strange enough, they will see all the other people in the paradise too).

Indeed, these conclusions are evident. According to the quantum mechanics (in its many-worlds interpretation) all feasible alternative classical states of the world coexist. Therefore, in future those states of the world which were called hell and paradise will coexist (as parallel worlds). In each of these two parallel worlds all people will be present as observers (in the first case they will be lost together with the whole mankind). However, if we consider the subjective perception of each of the people, the result will depend on his/her consciousness.

Let us consider the person who changed his/her consciousness into ecological one. The ardent desire of such a person is to see around him/her people with the same type of consciousness and resulting florescence of life at Earth. Because of this ardent desire (accompanied by the belief that it will fulfill) the subjective probability will be great for this person to see the corresponding destiny of the world. He therefore will perceive the future

---

[2]Actually it is not necessary that the consciousness of every human change properly. It is quite enough if the consciousness of most people change.

when the most people possess the ecological consciousness, high technology is not used for evil, and the global technological crisis is avoided. Of course, this person will see in this "paradise" world all people of Earth.

Analogously if another person does not change his/her egotistic consciousness, he/she will hardly believe that most people on Earth will do this. Therefore, for such a person the subjective probability will be high that he will find himself in the "hell" type of world. Of course he will see that all people perish with him/her and all living beings.

All this resembles the picture of the *Last Judgment*. Necessary difference is connected with the diversity between the Alterverse (set of parallel classical worlds, or alternative classical realities) and the conventional world with a single classical reality.

9.1.5　*Life sphere: making the concept more precise*

It seems that the contradiction appears in the very definition of the *sphere of life*: if there is a version of the loss of the entire world, then how this version is consistent with the concept of the sphere of life? There are two answers to this question which in fact make more precise what does the concept of life sphere means. It is convenient to formulate these answers in religious terms:

- God gives ability to select (the subjectively observed reality), but it is a human being who does make the selection. It can select good or evil. God ensures only that the selecting good leads to the life, while the choice of evil may produce death. The concept of life sphere is therefore equivalent to the concept of God. Blindly subordinated to God (that is characteristic for animals) provides being in the sphere of life, but the judgment about the good and the evil made by human beings can be erroneously, and behavior based on this judgment can go out of the sphere of the life (see Sect. 9.1.6).
- In Sect. 9.2 life of soul after death of body is described, paradise being the final goal of existence in "the other world". Applying similar considerations to the life after the global crisis, we may come to the conclusion that those, who selected evil, will perish in the global crisis, but already after death of these people, their souls will finally get into paradise. This is not paradise on the Earth for the living people, but paradise for the souls of righteous men and cleaned sinners.

9.1.6   *The Fall and the tree of knowledge*

The above consideration suggests an interesting interpretation of the *Fall*. The interpretation accepted by most people is that Adam and Eve accomplished the carnal sin. However, But if so, then what does here mean the tree of knowledge? Mentioning the tree of knowledge in this context was always strange for me.

Let us interpret the Fall in another way instead. The sin of Adam is that, after tasting from the tree of knowledge, man learned, that he can be similar to God, particularly create miracles, manage reality. He can not be simply subordinated to God, but he may solve by himself, which is good and which is bad. But man cannot know precisely what is good, because he cannot know about the distant consequences of one or another selection. While Adam was blindly subordinated to God, there was the peace on the Earth. When he began to select his solutions by himself, the evil settled on the Earth.

Evil is brought into the world not by God (i.r., it is not a consequence of the principle of life). It follows from the incorrect understanding by man, what is good and what is evil, and, because of this, incorrect selection of the desired scenario.

It is interesting that in this parable, when man tasted from the tree of knowledge and began to select his own solutions, evil settled also among animals, they began to eat each other. This is easily explained only within the framework of the concept of parallel worlds. It is difficult to explain, why a change in the system of the thoughts of man changed the behavior of animals. But it is easy to explain what man observes around him, if we assume that he fells into that of the parallel worlds, in which the evil rules, including among animals.

9.2   Soul and life after death of body

Is it possible, within the framework of the Quantum Concept of Consciousness (QCC) and Quantum Concept of Life (QCL), to interpret such concepts as paradise, hell and purgatory? If this is possible, then, probably, it is necessary to interpret also the concept "soul", which is closely connected with the idea that "the life after death" is possible in some way or another.

Let us examine these questions, yet realizing that the consideration is more speculative in this case than in case of the basic points of the quantum concept of life, which have been discussed until now.

9.2.1 *Soul before and after death of the body*

What does remain after death of a human being? His body is disintegrated and converted into the heap of molecules, which cannot be named a living being. Consciousness of course disappears, but what about unconscious?

At the fist glance, the question is meaningless, and this is actually has no sense if unconscious is treated only as the state of a human that is opposite to the conscious state. However, in the framework of QCL unconscious has a wider sense, as being in the quantum world as the whole, having access (in some way that hardly can be characterized quite precisely) to all parallel classical worlds forming this quantum world. The dead body is of course existing as an object of the quantum world, so the question about the access to all classical parallel worlds is not necessarily meaningless.

If we consider the very moment of death, or rather the period of gradual fading of life in the body, then it is clear that the state of unconscious and access to the parallel worlds remain during this time. The concept of "being unconscious" becomes meaningless after the "complete death" (although this is evidently not well defined term), but it is not evident whether the concept "access to the parallel worlds" remains meaningful or not.

Why does appear doubt about the fact that the concept "access to the parallel worlds" preserves its sense after death of body? Because it becomes unclear, who "has the access", if the man died. However, if one analyze more closely the question of "who has the access", the answer turns out to be not evident also before the man died. If the consciousness of a living person is switched off and he is in the state of unconscious (in the state of sleep or trance), then who has an access to the parallel worlds and what is meant by the expression "access to the parallel worlds"?

During the construction of the quantum concept of consciousness we avoided this question. The result of the process was the only important for us. The question about the nature of substances or objects participating in this process we laid aside on purpose. Just this strategy led to the success, allowing to overcome limitations that are usually arise because of the too specific understanding of materialism.

Now we also may remain on this position and speak about "the access to the parallel worlds" (both before death and after it) without mentioning explicitly, what substance has this access. However, for simplicity of terminology we may introduce a new concept, which designates the "substance" which has an access to the parallel worlds, when the consciousness of man is switched off. Let us name this concept "soul of man". Then we may say that when the consciousness of man being turned-off, his soul obtains access to the parallel worlds. This makes intelligent the assertion that the same substance, soul, preserves this access even when the man already died. For this being meaningful, it is sufficient to assume that soul of man continues to exist (we do not refine, in what sense) even after his death.

9.2.1.1   *Soul after death: judging the life*

What does form the essence of life according to the Quantum Concept of Life? This is existence in the quantum world, but in the restricted set of only *favorable scenarios*, which is called the *sphere of life*. That "something" which possesses the ability to live, understood precisely in this sense, is a soul. The body also possesses the ability of life, but understood in another sense, as the proper functioning of its organs.

If we accept this statement about the connection of the concept of life with the concept of soul, then it is possible to correctly raise the question, what happens with the soul after death of body. Indeed, selecting one set of scenarios or another has definite meaning even if the body have died. Let us try to make use of the idea of life of soul (after death of the body) to interpret the religious terms of paradise, hell and purgatory. For this aim we have to consider the stages which the soul passes after death of body.[3]

9.2.2   *Estimate of life criteria and judgment on the spent life*

After the death of the body special role is played by the set of scenarios, in which all people are guided by the same collection of the *criteria of quality of life*, as the dead person was guided in his life. Making use of the religious terminology, we can say that the soul makes use of this special set of scenarios in order to estimate the life criteria elaborated by the dead may during his life. We shall argue that this set of scenarios looks (for the soul of the dead man) as the paradise if the died man was righteous, looks as the hell for the sinner one, and looks as the purgatory in the general case. This gives the estimate of the life spent by the dead man.

But why the soul needs any estimate of the dead man? It is because the soul needs to select criteria of life for her eternal existence.

---

[3]Yet we remember that all these considerations are more speculative that the main points of the quantum concept of consciousness. They may have some interest as the preliminary attempts to interpret the known religious concepts in scientific terms.

The soul tests various life criteria to find such set of them, which make her eternal existence comfortable. Testing any given set of life criteria is a stay in such world, in which all people are guided by precisely this set of criteria. The soul finds out, whether it is comfortable to live among people, which are guided by these criteria. Of course the soul begins testing that set of criteria, which has been elaborated in the course of the life, because it was, apparently, selected by the man as optimum.

A stay of the soul in the world thus appearing, turns out actually to be the judgment on the spent life. Estimation is thus given to this life (or to this personality), and stay in this world turns out to be reward for the (soul of) righteous life or punishment for the (soul of) culpable life.

9.2.3   *Estimate of life criteria - more details*

Consciousness is the ability of a living human. This ability no longer appears after final death, but its disappearance may come not immediately. Apparently, the consciousness may remain actual in the transition period, becoming gradually weaker. Let us recall that consciousness, according to the quantum concept of it, is the separation of parallel worlds (classical realities). This means that in the transition period the possibility remains for the dying man to subjectively perceive one of parallel worlds or a restricted set of the parallel worlds independently of the rest (the separation of alternatives yet did not disappear completely). The soul was not yet completely detached away from the material (visible) world.

However, the separation of alternatives is weakened; therefore it is possible to subjectively experience not a single scenario, but the wider set of scenarios. It is important that, since the consciousness is connected with the emotions, the partial retention of consciousness makes it possible to perceive stay in the restricted set of scenarios emotionally, experiencing bliss or sufferings. This is necessary to interpret the above mentioned state in terms of paradise and hell.

Thus, in the period of death and immediately after it the soul is partially freed from the connections, which she had with the life of body. But, apparently, within the sphere of life soul can select the niche, in which she desires to exist. In order to make this selection, soul investigates various scenarios. In this study, soul can make use of her experience during the life of the body, but she can also to look into the future.

Glimpsing into the future, soul sees that it is necessary to remain in the sphere of life, in the sphere of survival. But besides this general conclusion it is necessary to make a more concrete specific selection, the selection of such niche (such subset of scenarios), in which the quality of life is sufficiently high. It is necessary to remain in the narrower sphere of life of high quality.

But what does it mean - the high quality of life? By what criteria one should judge of the quality of life? Soul can use that collection of criteria, which the dead man had in his life. However, the soul must test this collection, ascertain that it is actually good, what it seemed during the life.

To realize the test of the collection of the life criteria, the soul immerses herself in the specially selected subset of scenarios. Each scenario in this subset is the evolution of a world in which all people are guided by the same collection of the life criteria, that the dead man had in his life. In order to obtain the emotional characteristic of this collection, soul uses the possibility to partially include consciousness i.e., to experience (also emotionally) the given subset of scenarios independently of the rest. Situation appears, which can be named paradise, hell or purgatory.

Indeed, if the criteria of the dead person were noble, the soul of this person will be surrounded by noble people in the world created for the testing. She will experience happiness, feeling herself in paradise. But if the dead person was a large sinner, now all people around his soul in the testing world will prove to be the same poor people. The soul will feel very bad, this there will be hell or purgatory for her. The experience of a stay in this world will give for the soul the possibility to judge, in what respect the criteria elaborated by the dead person are in fact not optimum. After improving these criteria, the soul will form the world, populated by righteous men, i.e., finally she will fall into the paradise.

Thus, to estimate the quality of those life criteria, which the dead person had in his life, his soul examines the set of scenarios, in which all people rest on these criteria. The sensation of such scenarios is paradise, hell or purgatory. In case if the dead person was righteous man, then the sensation from this set of scenarios occurs paradise, because all people in it are righteous men and the world is benevolent. If the died man was a sinner, then all people in this set of scenarios are sinners and the resulting world is unfavorable.

Passing this stage, soul gets to know the true value of those criteria, which were characteristic of dead person. If the criteria were universal, ecological, then the surrounding world is favorable and the soul experiences bliss. If the criteria were selfish, then the world is hostile and the soul experiences sufferings.

Improving the criteria on the basis of this experience, the soul finally remains in that subset of the scenarios, which is determined by the universal, ecological criteria. She understands after the experience of purgatory, what criteria lead to the bliss, and she remains in the sphere, determined by these criteria. She finally settles into paradise and experienced eternal bliss.

9.3 Karma and reincarnations

The previous reasoning followed a certain logic, but the only concept from the spiritual practice, which was used in this consideration, was the concept of soul and life of a soul after death of the body. Now we will see, that the conclusion achieved in a purely logical way can be interpreted as describing the ideas of Hinduism and *Buddhism* about karma and reincarnations.

Let us ask ourselves, what does it means that the soul, after death of the body, investigates the world, which is arranged according to those vital rules (criteria of life), that were developed by the dead person. This means that the soul investigates narrow set of classical realities and their evolution in time. Somewhat simplifying, we can say that the soul investigates a single classical reality and its evolution, i.e., a single classical world.

But this is nothing else than the definition of the *subjective* perception of the quantum world, because subjectively only one of the parallel (Everett' s) worlds is perceived. In other words, the soul experiences new earthly embodiment, reincarnation, but this time the world, in which she is personified, its quality, depends on what *criteria of life quality* the possessor of this soul developed in the previous life. The more noble were these principles, the better (with respect to this person) is that world, in which the soul is personified this time.

This exactly corresponds to the Buddhist's concept of *karma*. From what is the karma of the man in his past life, it depends, to what extent favorable will be the conditions for his next life. And from the fact whether he will improve his karma in the new life, his existence in the next embodiment will depend. Experiencing long series of reincarnations, the man can be completely purged of sin, achieve enlightenment and taste nirvana, i.e., infinite bliss. Then his soul will not experience the needs for the new terrestrial embodiment and he will remain in "the other world" (in our terminology, he will be permanently existing in the quantum world, i.e., will always have an access to the entire set of parallel worlds).

The carried out reasoning shows that, from the point of view Of Quantum Concept of Life, the picture of life after death given in Buddhism (the long series of earthly embodiments leading to the enlightenment and the nirvana), is nearer to the "truth" than that, which Christianity gives (the only judicum dei with directing to the paradise or hell).

PART 5

Summing up the results

In this part we will summarize the results, which were obtained in this book. The brief presentation will be given of the logical scheme of the Quantum Concept of Consciousness (QCC) and Quantum Concept of Life (QCL), as well as the main consequences, which follow from these concepts. In the conclusion we will comment on the obtained results from various points of view, some of which are important for physicists and the other which have an interest for the wider circle of the readers. In particular, we will discuss the philosophical aspects of the obtained theory. Tracing the history of quantum mechanics in the 20th century and at the beginning of 21st century, we will justify the conclusion that working on quantum theory of consciousness and life realize the final stage of the scientific revolution, begun by the creation of quantum mechanics.

Chapter 10

Main points of the Quantum Concept of Life (QCL)

In this chapter we will very briefly present the logic circuit, which leads to the Quantum Concept of Consciousness (QCC) and Quantum Concept of Life (QCL), and some consequences of this concept. Purpose of this is in compact surveying of the constructed theory as "from the height of bird flight". One of the tasks, which thus is solved, is demonstration of the fact that the logic of the QCL is very simple and can be brought to the compact scheme, which makes this concept more plausible.

Nevertheless, the more reliable proof of its correctness is the set of consequences of this theory, which, as it occurs, makes it possible to explain many of the unexplained phenomena.

10.1 Logical scheme of the quantum concept of life

Let us present explicitly the chain of logical steps leading to the conception of consciousness and life developed above (see Fig. 10.1). We shall start with the logic of the quantum mechanics itself (in the version given to it by Everett), then go over to the minimum extension of it leading to the conclusion about the *super-intuition*, then to further extension giving also probabilistic miracles, and finally expose the concept of life in terms of (Everett's) scenarios.

10.1.1 *Quantum reality*

1 (statement of quantum mechanics). Objectively there exist parallel (Everett's) worlds.

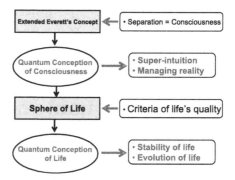

Fig. 10.1   Logical chain from quantum mechanics to theory of life

Quantum-mechanical formalism, namely, the linearity of the evolution of quantum system, predetermines, that objectively there exist parallel worlds, or alternative classical realities (shortly, alternatives). This is the essence of quantum reality in contrast to classical reality on which our intuition is based. The surrounding us world, in which quantum reality rules, may be named *Alterverse* (the quantum world = the set of alternative classical worlds = Alterverse).

For the first time this issue is formulated within the framework the interpretation of quantum mechanics, proposed by Everett. However, rejection of this assumption makes quantum mechanics not logically closed, and in this sense the adoption of this assumption is unavoidable.

Objection against this lies in the fact that on the experience we never see many worlds, we always see only one world.

However, what in this is surprising? If it is asserted that parallel worlds objectively exist, this does not mean that subjectively we must perceive all these worlds. This conclusion can be drawn only if the objective is identified with the subjective, that the objectively existing reality is identified with its subjectively observed image. However, this identification is by no means compulsory.

2 (Definition of the subjective) Consciousness   separates   parallel worlds, so that in the subjective perception an illusion is created, that there is only one world.

This point explains, why we subjectively do not perceive existence of parallel worlds.

3 (Consequence) In principle it is possible to obtain information from all objectively existing parallel worlds and reproduction of this information in the consciousness, i.e., in the subjective perception.

If objectively there exist parallel worlds, then in principle obtaining information from all these worlds (i.e., from the quantum world as a whole) and even reproduction of this information in the subjective sphere is possible. Since the evolution of the quantum world is reversible in time, information from any point of space-time is accessible.

10.1.2 *Quantum Concept of Consciousness (QCC)*

The question arises, how is it possible to obtain information from the objectively existing parallel worlds and to accept this information subjectively (by consciousness). Answer to this question follows from the first assumption of QCC:

4 (Assumption in QCC) Consciousness is separation of the alternatives (parallel worlds).

This assumption simplifies the logical structure of quantum mechanics, since instead of two primary concepts ("consciousness" and "separation of alternatives") only one remains. Besides this this single primary concept is now characterized from two qualitatively different points of view: from the side of quantum mechanics and from the side of psychology.

5 (Consequence) The super-intuition, which draws information from the parallel worlds, arises in the state of unconscious.

Indeed, if consciousness is the separation of alternatives (parallel worlds), then the turning-off the consciousness removes this separation and opens access to all alternatives.

Already at this level the new consequences appear, which make it possible to explain strange, not explained otherwise, the abilities of consciousness, namely, super-intuition, in particular, scientific insight, clairvoyance, in certain cases the soothsaying, or fortune-telling.

This is the minimum scheme of the quantum concept of consciousness. It include only one new assumption, which is especially plausible because it simplifies the logical structure of theory and leads to the important consequences.

10.1.3 Quantum Concept of Life (QCL)

6 (Assumption in QCL) A man (and a living being generally) can influence the subjective probabilities of the alternatives, increasing probability to experience those alternatives that are favorable.

This assumption is arbitrary. However, it is more plausible in light of the fact that life, when it has access to some information, then it always develops the means of the usage of this information to improve the quality of life.

This assumption makes it possible to even more enlarge the spectrum of predictions and the spectrum of phenomena explained on this basis.

But, it goes without saying, more valid reason for accepting this assumption is the fact that actually are observed the probabilistic miracles, whose possibility follows from this assumption.

7 (Consequence) Probabilistic miracles.

This completes passage to the Quantum Concept of Consciousness (QCC) and even generally Quantum Concept of Life (QCL). Consequence of this concept is the explanation not only of consciousness, but also of the phenomenon of life. Furthermore, the classical nature of alternatives is explained.

8 (consequence) Nature of the phenomenon of life.
9 (consequence) The classical nature of alternatives.

10.1.4 Quantum Concept of Life (QCL) in terms of scenarios (sphere of life and the principle of life)

The Quantum Concept of Life thus appearing can be formulated without the usage of the concepts of consciousnesses and unconscious (which are characteristic only for humans beings and may be highest animals). For this aim we assume that the perceived by the living beings evolution of the world (including the evolution of the bodies of themselves living beings) is presented not by the totality of all possible (from the point of view of natural sciences) scenarios, but by a narrower set of scenarios, favorable for the life. This narrower subset of scenarios can be named the sphere of life.

Formulation of QCL The *principle of life*: The perception of the evolution of life is presented by the subset of favorable scenarios (forming the *sphere of life*).

Commentary The principle of life is analogous to the antropic principle.

The *principle of life* in its structure is analogous to the *antropic principle*. The life principle is in essence a version of the antropic principle, but with 1) Homo Sapiens as an observer replaced by the totality of all living beings and 2) *Multiverse* (the set of many universes existing besides with each other) replaced by *Alterverse* (the set of the virtual classical worlds presenting a single quantum world existing in the sense of quantum reality).

10.1.5   *The extended scientific methodology must include the subjective*

The Quantum Concept of Consciousness and Quantum Concept of Life require the expansion of the scientific methodology. Besides the objective, scientific methodology must include the subjective: not only measurements and their results, but also the images of these results appearing in consciousness. The situation appears that is very uncommon for the science from the point of view of methodology. Specifically, in some situations it is not possible to distinguish the objective from the subjective. In particular, probabilistic miracles cannot be distinguished for the random events, which are inevitable in quantum mechanics.

Any probabilistic law can be refuted only by an infinite series of the observations, in each of which this law is broken. Any final series of observations, contradicting to this law, in reality does not refute this law. In this sense the random events, which subjectively look like probabilistic miracles, are completely compatible with the objective laws of quantum mechanics.

10.2   Consequences

In the above chapters many consequences of the Quantum Concept of Consciousness and Quantum Concept of Life have been mentioned. The task of the present section is to give short illustrations for those who wish to obtain at least the first idea of them without reading long texts.

We shall present here only two main classes of consequences, super-intuition and probabilistic miracles, providing some examples of them.

10.2.1   *Super-intuition*

Super-intuition appears as a result of the access to the entirety of parallel worlds and ability to obtain information from there. This ability appears in the state of unconscious, during the disconnection of consciousness. The information thus obtained is unavailable in the conscious state, when (according to the subjective sensation) only one of these worlds is perceived.

Information from the parallel worlds (or a certain part of it) remains upon returning to the conscious state and only in this stage takes the customary form of thoughts or visual images. The initial form of this information is entirely different, it is not expressed by conventional means. Therefore the appearance of this information before the mental look of the man in the customary means is experienced as and "enlightenment", i.e., the sudden and "unexplainable" appearance of new knowledge "from nowhere".

We can name *super-cognition* the regime of unconscious, in which extracting information from the parallel worlds occurs. By the term of *super-consciousness* we shall designate the process, in which the person, who is in a state of consciousness, obtains information from the parallel worlds (super-information) with the help of temporarily sinking into the unconscious to make use at the process of super-cognition.

One may transparently illustrate the process of super-consciousness if presenting the unconscious (the invisible world) as a see and the conscious (the visible world) as the air above this see. Then the usual state of consciousness is the flying in the air of the conscious (in the visible world). Diving into the see of unconscious (the invisible world) enables obtaining super-information in the process of super-cognition. Super-consciousness is the ability to swim on the surface of the unconscious (between the conscious and the unconscious) with periodic temporary diving into the unconscious (which makes it possible to use the process of super-cognition) and bringing the obtained super-information back into consciousness. Super-consciousness is lifting super-information from the unconscious to the consciousness.

10.2.1.1   *Clairvoyance and soothsaying*

clairvoyance is the power or faculty of discerning objects not present to the senses (i.e., information unavailable in the conscious regime).

The set of all parallel classical (Everett's) worlds is nothing else but a single quantum world. But according to quantum mechanics the evolution of quantum world is reversible and therefore its state at any moment determines its state at all other moments of time in the past and future. Therefore for the quantum world (considered as a whole) there is no sense in the concepts of the past, future and present, and all space-time points are equal in the quantum world.

Therefore the process of super-cognition makes it possible to extract information from any space-time region of the quantum world. In simpler wording, any, arbitrarily remote spatial domains of all parallel world, and moreover the state of these regions at any times in the past and future, become accessible in this process.

This ability of consciousness is manifested in the life as clairvoyance. One of the most known clairvoyant was Edgar Casey, who helped hopeless patients, obtaining with the aid of the clairvoyance information about which means would provide their recovery (see Sect. 2.2.1). One more example of a very strong clairvoyant is Wolf Messing, who lived in the first half of the 20th century in Europe (first in Poland and Germany, then in the Soviet Union). He could, in particular, read thoughts and learn about the fate of a man from his photograph.

Both Casey and Messing sometimes practiced *soothsaying*, or *fortune-telling*. In particular, they predicted the beginning of the Second World War. In connection with this let us note that, according to the Quantum Concept of Consciousness, the prediction of future in principle cannot be quite reliable. Indeed, the future is represented by various alternatives. If one has to predict which of these alternatives will be perceived subjectively, he can do it only with the specific probability,but not with certainty. A prediction may become practically reliable only if one deals with the not too distant future and predicts such an event, which "is prepared" already by the course of events in the past and present. In the case of the Second World War precisely this occurred: it is known now that for several years prior to its began the war was actually inevitable.

Consider now the case of predicting insignificant events, which can with sufficient probabilities either occur or not occur. we can formulate the following two remarks about predictions of such events.

First, more reliable in this case are conditional predictions, when it is predicted, that the event will occur under the condition that another (definitely specified) event will occur. In this case the prediction contains information not about the event itself, but about the necessary connection of the two events with each other. Such connection may be quite determined, and then the prediction is quite reliable. the examples of such conditional predictions are "readings" of Edgar Casey, who predicted, that the certain patient will recover if the specific drugs will be given to him or the specific procedures will be performed with him. Of just the same nature is the prediction Of Messing that Hitler is defeated, if his impact will be directed to the east.

Secondly, in case of unconventional predictions of "ordinary" (not world-scale) events one may think that the clairvoyant does not simply predict these events, but predetermines them. Let, say, a certain event in the future can occur or not occur (for example, patient can recover or not). This means that two corresponding alternative realities will appear in the future. Let us assume that the clairvoyant predicts that the event in question will occur (the patient will recover). Thus he creates "information bridge" between one of the alternatives in the future and that of alternative realities in the present, which is subjectively received. After this, those, who perceive now this reality (those, who hear this prediction the clairvoyant or those who can in principle know about this prediction), will compulsorily perceive in the future just that alternative reality, which has been forecasted.[1]

In this case fortune-telling results not in the prediction of something that has to occur anyway, but rather fixing one of all possible alternatives in the future. All the other alternatives are then excluded for the subjective experience of people (at least those who know or in principle can know about the result of the fortune-telling).[2] It is clear then why fortune-telling is sometimes estimated as dangerous. People who possess this ability (just as the ability to produce probabilistic miracles) have to use it only as a last resort. By the way, Messing confessed that he did not know whether his talent was a gift or an anathema. This is in agreement with the considerations of Sect. 9.1.6.

Edgar Cayce and Wolf Messing were examples of great clairvoyants. But each of us can see less impressive examples of clairvoyance in the ordinary life, if he/she does not have prejudice against this. Frequently the clairvoyance is manifested as super-intuition or presentiment.

In the special circumstances the individual cases of clairvoyance occur with ordinary people, which do not show any special abilities in their everyday life. They possess nevertheless the potential ability of clairvoyance which may become evident at the dramatic moments, for example, when danger threatens to this person or someone of his close relatives.

---

[1]let us note that in another alternative in the present the same clairvoyant predicts another alternative in the future, and this future alternative will be fixed for those, who hear this another prediction in the present.

[2]Of course, we assume that the prediction is performed by a genuine clairvoyant, not a charlatan.

10.2.1.2 *Scientific insights*

Super-intuition explains such a well-known, but yet mysterious phenomenon as a scientific illumination (*insight*), i.e., instantaneous guess about how to solve the long ago confronting problem. Such an enlightenment unexpectedly comes as if "from nowhere". It is not based at rational analysis of existing information and frequently indicates on the qualitatively new way of thinking, the new paradigm, which earlier not at all figured in the analysis.

Sometimes scientific illumination contradicts the existing information, but, it is however strange, finally erroneous proves to be not the irrationally found guess, but just the existing information. The explanation of such direct viewing of true solution of the serious problem may be (not necessarily always) in that the super-consciousness traces the consequences of each of the candidate solutions of the problem and reveals which of these solutions is confirmed in the future. In this case the super-intuitive solution of the problem is a sort of foresight.

10.2.1.3 *Efficient method for solving problems*

Many great scientists (among them Einstein) gave evidences of the scientific insights. However, the scientists of not such high rank also use them (sometimes not being aware of this) in their work. This fact explains one paradoxical element of the *efficient methodology of the scientific work*: at the moment when, after long rational work, the key decision should be made (either about the conclusion from this work or about further direction of the work), the scientist must temporarily stop working on this problem instead of obstinate continuing working on it. Then consciousness is turned off this problem so that the problem is in the sphere of unconscious and is worked out with the help of super-cognition.

It is paradoxical, but this procedure for solving problems gives fantastic effectiveness, is in practice easier than that which commonly is used. What does the researcher do when he in his work meets a serious obstacle that make impossible for him to move further? He can, of course, stop further work, deciding that the problem confronting him has no solution or is too hard personally for him. But if he is sufficiently persistent, then he continues to work, only increasing his efforts, may be sitting by nights, wasting heaps of paper but again and again suffering failure in his attempts to solve the task.

But in reality these agonizing attempts are not necessary. One may do in entirely different way. He has to accurately formulate the state, which appeared at this stage of solving the problem, noting all partial results obtained, the questions not solved, and, as far as possible, the reasons, which do not make it possible to move further. After this, one has to forget about this problem, completely shake it out of his head and to turn himself to relaxation. It is important in this case to preserve confidence that the problem will be finally solved.

If all this is done, then after a certain time, completely unexpectedly, at the moment when there is no even thought about this problem, arrives the guess about what must be done in order to find the solution. This can be a scheme of solution or guess about a new way which makes it possible to move further. This makes it possible to resume regular work and to realize the arrived guess. The problem will be solved in a certain time or the essential advance in its solving will be achieved. If again, already at another level, a fundamental difficulty arises, the same method should be applied: clearly formulate the state achieved in the new stage of the solution, then forget about the problem and return to it again only when a guess arrives about the next step in the solution.

In this way complex problems must be treated, those which cannot be solved "in a single course". Each of the described stages plays the role of the loop of feedback. The "guess" found with the aid of the super-intuition makes it possible to return at the beginning of the process of the solution, but to pose the problem in a new way, understanding it more deeply or with the more suitable means of the solution.

Some researchers will object, indicating that they successfully solve problems exactly with the aid of persistent, continuous, sometimes agonizing working on them. According to their experience, the guess about the correct method of solution comes just in the course of continuous working. Does this mean that the above described procedure is not applicable for these scientists? Completely no.

It is possible that these people actually apply "usual" methods, not including super-intuition. Of course this can lead to the solution, especially if sufficiently simple problems are under solution. But the different version is also possible. It is possible that the turning to super-intuition nevertheless occurs also in the case of continuous working, but at a heavier cost.

Let the man is sitting at the table entire night and can obtain no result, but toward the morning a guess appears about the new way that can lead to complete or partial solution.[3] It is strange, but this can be the result of super-intuition, which appears, as always, with turning off the consciousness from the problem. However, turning off occurs in this case not at will of scientist, but because he got tired and, as a result of fatigue, every once in a while for short periods either ceases work or even continues to work "automatically", performing customary routine operations without concentrating attention in them.

One may advise to this scientist to apply the above described procedure, i.e., to cease the work on the problem at the key stages, expecting super-intuitive guess. The results will be not more badly, most likely better, but efforts spent for their achievement will be much less.

When I told about this to one of my colleagues, he asked: "Did you yourself tried to apply this procedure?", and immediately continued the conversation, without waiting for an answer. He was confident that the answer should be negative. But in reality I did use and am using this procedure, and results are magnificent.

This began many years ago, when I was working in Germany, in the University of Konstanz. I learned about this procedure of the solution of problems completely unexpectedly, from the source, where I never expected to find councils for scientific work. This was the book of Ramacharaka "Rajah-yogi" in Russian translation. The discussion in in this book was of course not about science. But the procedure for solving problems which had been presented in this book was universal, and I applied it to those physical tasks that stood before me. The results were splendid.

I mastered this procedure approximately two years. At the end of this period I knew definitely about each task, which appeared before me: if it has a solution, I will solve it. Since all problems, one after another, were solved smoothly and easily, I may suppose that the very selection of tasks (i.e., judgment about whether a certain task can be solved or not) also came as a result of super-intuition. Finally I so mastered this procedure that no longer thought about it.

10.2.1.4 *Chess*

Super-intuition is well-illustrated by the example of playing chess. Of course one can use calculation in this game, sorting out the versions of his

---

[3]Many people will confirm that this is typical situation. Actually, in case of night working the guess frequently appears already in the middle of night or toward the morning. In reality this occurs in the specific hours, which are on some reason favorable for the intuition, which by itself is extremely interesting.

motions and motions of his opponent and evaluating the result of each version. Specifically, motions are selected in this way, if not living person, but computer plays chess. All versions thus considered form a tree, in which the selection of sequential motion corresponds to several branches, which grow in the same point.

The difficulty with the tactics of sorting versions is that for evaluating the correctness of one motion it is necessary to examine all versions of the motions, which follow after it, and the number of versions very rapidly (according to the power law) grows with an increase in the number of the examined motions (depth of calculation). In order to decrease the number of the analyzed versions, the previous experience or the theory of chess game is used. This makes it possible to cut off some branches of the tree already at the level of the first motions. This seriously reduces the number of versions to be examined; yet, it nevertheless remains very large.

For a computer no other tactics, except sorting versions, there exists. However, a living chess player in addition to this tactics can use the ability of super-intuition. For this aid, at the key moment of considering the motions he must turn his consciousness from this task so that its solution would continue in the regime of unconscious. In this regime sorting all versions of the game at any depth is performed. This supplies the information about what chain of motions leads to the victory. Returning to consciousness, the chess player realizes, unexpectedly and without any rational arguments, what motion he has to select and, possibly, what motions must be selected after various backlashes of his opponent.

Modern supers-computers play at the level of grand chess masters, and their improvement occurs very rapidly. Therefore if human chess players calculate their motions via the sorting of versions, then already within the next few years computers will invariably conquer them. But if chess players in reality use super-intuition, then situation is entirely different. In this case with any increase in the power of computers will appear such brilliant chess players, who will beat them.

Invariable victory of human beings over computers will cease if and only if the power of computers will be sufficient for the total calculation of the chess game. But in this case this game will generally lose sense, because the white figures, that begin chess game, will always win or at least attain no one's.

10.2.1.5   *Is artificial intellect feasible?*

The robotization of chess game is the very obvious case of the more

general common direction in the science and technology, which is traditionally called the creation of the *artificial intellect*. This direction appeared
simultaneously with the advent of the first computers and was based on
the idea, that the human intellect may be reduced to the work of the brain,
which works approximately just as computer differing from it only by more
sophisticated construction.

**History of the problem: from computers to "quantum conscious-
ness"**   A question about the feasibility of an artificial intellect was then
reduced to the question whether it is in principle possible to create a computer, that could solve the problems, which the human brain solves. In this
formulation the question could have only one "obvious" answer, which, if
we omit details, consisted of the following. The brain is a certain material
system, and there are no obstacles whatever in order to create another material system, whose functioning exactly simulates the functioning of the
brain.

It was discovered at some stage that hierarchical architecture of the
known (by that time) computers principally differs from the structure of
the brain which is essentially the network of cells, neurons, connected with
each other.

This discovery had, as its consequence, a certain breakthrough in computer technology: the creation of computers having architecture of a new
type, similar to the network of neurons. Such devices were called neuro-
computers. They could more efficiently than usual computers solve some
computational problems.

However, this did not change principally how the problem of the creation
of the artificial intellect was understood. Unfortunately, even now some
specialists understand it in exactly the same manner.

If we are based on the Quantum Concept of Consciousness, then the
problem of the artificial intellect is represented completely differently. The
reason is in the fact that, according to this concept, the problems met by
a human being can be solved by two principally different methods.

Some problems can be solved with the aid of a certain sequence of
computational (in particular, logical) procedures. They are solved by the
brain, which works in this case as a computer.

However, there is another class of the problems confronting human.
These are the problems which in principle cannot be solved with the aid of
the computational procedures, if one is based on the information accessible

in the conscious state.[4] A man has at his disposal another method of solving such problems: super-intuition, or super-consciousness, i.e., sinking into the state of unconscious, when access appears to the entire set of alternative worlds (realities), obtaining super-information and its (or its part) reproduction in the state of consciousness. With the presence of this information the problem can be solved by conventional computational means, i.e., due to the functioning of the brain. But in many cases super-information contains already final solution of the problem.

How the brain (considered as a computer) is arranged is not thus far completely known. However, in principle it can be found with any degree of detailing and after this the computational work of the brain may be simulated, most likely on another material basis. As a result such a device can be created that actually may be named an artificial intellect; however, super-intuition, or super-consciousness that is characteristic of man, can be realized by no technical equipment.

The final conclusion may be formulated as follows: it is possible to create an artificial intellect, but it is impossible to create artificial living being.[5] A difference of the living chess player from a robot of Sect. 10.2.1.4 illustrates this conclusion.

What can quantum computer do?   Quantum computer in the usual sense of this term is an information processing device working in the quantum-coherent regime. For realizing this regime, the set of the degrees of freedom (qubits) included in the information processing should be strictly isolated from its environment. This is the main difficulty for realizing quantum computers (although the requirement of isolation may be weakened by means of the error-correcting codes).

For readout of the computing results, after the necessary cycle of unitary evolution of the computer, some observables of this quantum system undergo measurement. This causes decoherence of the quantum system and brings the results of computing process into classical form (which may be stored as long as is necessary).

Unlike classical computer, quantum computer can be used for solving only restricted number of problems, but with much greater speed (because of quantum parallelism, i.e. possibility to parallely process enormous

---

[4]It is shown in the works by Penrose [Penrose (1991, 1994, 2004)] that a man can solve problems which cannot be reduced to the computational tasks.

[5]For simplicity we spoke of the super-intuition of a man, but any living being possesses the ability of super-cognition, which ensures survival and which also cannot be reproduced by technical equipment.

number of data). However, just as a classical computer, quantum computer is inanimate material system and cannot intuitively (super-consciously) acquire information from "other" classical alternatives (other Everett's worlds). Direct vision of truth, although based on quantum effects, is feasible only for living beings.

10.2.1.6 *"Miracle of life" as an analogue of the super-intuition*

Chess player, who uses super-intuition in order to examine an inconceivable quantity of scenarios of game, well illustrates the possibilities of super-intuition. However, entirely different functions, which do not refer to consciousness, can be achieved according to the same scheme.

The unusual possibilities, inherent in the ability to access to the parallel worlds, or parallel scenarios, are used by no means only by people. This is a property of life as such, its essence, the definition of life. Just because of these enormous abilities life, in all its manifestations, is truly miracle. Super-intuition appears while consciousness is turned off. Access to information about all possible scenarios not only does not require consciousness, but it is, on the contrary, achieved only in the state of unconscious. It is clear that all living beings can use such access, regardless of whether they do possess consciousness or not. And precisely this possibility ensures that which is called "the miracle of life" and which truly looks like miracle.

We will consider two examples, in which, apparently, is manifested the possibility of the sorting of the huge number of scenarios and selection best of them. First example is the capability for survival under the very wide spectrum of conditions. The second is the extremely effective evolution, which includes "inexplicable" jumps, transferring organisms or entire classes of organisms to a new level of complexity.

Survival The survival of living organisms is achieved due to their enormous capability for self-regulation. Somewhat simplifying picture, one may say that, after determining the surrounding situation with the aid of the sensory organs, the living being switches on a mechanism calculating the actions, which lead to the survival in this situation. However, it seems improbable that it is possible to correctly calculate effective behavior despite of great variety of the states of the environment and frequently unpredictable changes in this state.

The task simplifies, if it is possible to calculate how to behave during a short time, then to scan the environment again and again to design behavior for the following short period. However, such a strategy is in many cases unfit. Sufficiently frequently it is necessary to calculate optimal behavior for a very long period of time. This is necessarily if the best scenario of behavior for the long period (i.e., such, which leads to the survival during entire this period) is not the est in the short parts of this long time.

If this is the cases, then it is impossible to restrict the calculation of behavior by short time intervals. The selection of the scenarios, which are best in the short intervals, can lead to the loss in long time. In order to avoid this, it is necessary "to sacrifice quality" on some of the short intervals in order to win finally in long time. In such cases calculation should be made on the large time interval. But the complexity of the mechanism realizing such a calculation exponentially increases with the increase in the time interval.

For this reason it seems improbable that the effective survival can be ensured with the aid of the rational operations of the type of calculation. But the mechanism which is similar to super-intuition, easily solves the problem. Let us assume that the living being (having the consciousness disentangled from the problem or having no consciousness at all) obtains information about all possible parallel worlds at all successive moments of time. In other words, we suppose that there is access to the information about all possible scenarios of the evolution of the given organism and its environment. Then the information about what scenarios ends by surviving gives at the same time the plan of the behavior, which ensures survival. The length of the time interval in this case can be arbitrary because the access to the parallel worlds is based on the mechanism, which makes use of actual infinity.[6]

Health The described mechanism, which uses super-information from the alternative scenarios for surviving, determines, apparently, the very essence of the phenomenon of life, it is actually the definition of life. But in proportion to the complication of living organisms this mechanism begins to be used not only for the survival, but also for an improvement in the quality of life. For human beings this means, first of all, the maintenance of health at the sufficiently high level.

The maintenance of health means the guarantee of a constant state of the internal medium of organism (homeostasis) in spite of changes in the external conditions. This is achieved due to the action of special organs and regulating systems of the organism. These systems in turn obtain

---

[6]This is the special feature of the phenomenon of super-cognition (performing in the unconscious) differing it from every process or event dealt with in physics.

the appropriate orders from the brain, and also from bone marrow and, probably, from other main elements of nervous system. But as these orders are elaborated?

The information about how it is necessary to react to typical changes in the external conditions is of course part of the hereditary information. In other cases correct reactions are elaborated by means of the calculations, when the brain works as a kind of computer. However, it is obvious that the nonstandard situations, in which no computer can elaborate correct recommendations for the regulating systems of the organism, must regularly appear. Meanwhile the mechanism, based on the sorting of the alternative scenarios, is universal. It will give the correct answer in any case, and in any situation it will find out as the systems of the organism must react, in order to be adapted to the changed conditions.

Since the sorting of alternative scenarios is possible only in the state of unconscious, it is logically to assume that turning to parallel worlds and parallel scenarios occurs in the sleep, when consciousness is turned off. This is confirmed by many facts. In particular, in light of this assumption one may easily explain why "sleep cures", i.e., it proves to be beneficial for a person, who suffers any illness. Furthermore, this is explained, why a person, deprived of sleep, heavily falls ill and finally dies. This person does not obtain corrective information from the parallel worlds (from the quantum world) for too long, and the failures are accumulated in the organism, finally becoming critical.

However, many functions of regulation in the organism (temperature, blood pressure etc.) are carried out permanently, even in the state of wakefulness, but without the participation of consciousness. For the correction of such functions by means of the sorting of scenarios it is not necessary to disconnect consciousness, because the consciousness is always disconnected precisely from these functions. Therefore, turning to parallel worlds for the correction of some functions of the organism occurs constantly, while the other forms of correction are achieved in the sleep.

Simple organisms, which not at all possess consciousness in the usual sense of this word, have only the first, permanently acting mechanism of turning to quantum world.[7] With the complication of organisms and the

---

[7]Such organisms are nevertheless have the function of "reflection" of quantum world, which makes it possible to separate the alternative classical realities from each other. They live as if "on the border" of this separation, constantly obtaining information from all alternatives, but using this information in each of the alternatives separately from the rest.

appearance consciousness the second mechanism emerges, which acts due to periodic complete turning off of consciousness.[8]

From the free will and morphogenesis to jumps of the evolution

The possibility *to choose the alternative classical reality* easily explains many well known scientific facts, that cannot be explained by another method. We already mentioned the revitalizing force of sleep and its absolute necessity for the life. One may add to this explanation of freedom of will, morphogenesis (i.e., the process of gradual building the body of an embryo) and jumps of evolution. This is how one can argue to explain these phenomena.

It is not difficult to understand how a decision, made in some neuron, is brought to the muscles. However, how this decision does appear, i.e., how it may happen that from all versions of the neuron's state a single one is selected? According to the out Quantum Concept of Consciousness, no objective selection occurs All possible versions of the decision (neuron's state) appear and then realize in various parallel (Everett's) worlds. Super-consciousness can analyze the consequences of each of these solutions and increase the subjective probability of that world in which the best (from the point of view of its consequences) decision appears. Of this consists the freedom of will.

Let us go over to morphogenesis. In the genome the information is recorded about the final (after the morphogenesis is over) construction of the organism. But as it is being built step by step, beginning from one cell and to the extremely complex complete organism? In each of the intermediate stages, the consequences of the possible variants of the next step of construction are analyzed with the help of super-cognition, and the variant, which may, in the course of the following steps, lead to the correct (corresponding to the genetic information) construction of the organism, is selected.

Analogously by the work of super-cognition it is possible to explain the evolutionary jumps, which lead to a new structure of organism without any long chain of small changes, supported by the natural selection.

The well-known mechanism of evolution is reduced to random changes in the hereditary information (for example, due to the mutations) and subsequent selecting those changes, which proved to be favorable for life (natural

---

[8]They report about the people, which either sleep never or very rarely but preserve health. One may assume that,the health of these people is ensured only due to the first, evolutionarily more ancient, mechanism. This is a kind of atavism.

selection). This mechanism works only when the change occurs by means of small steps, each of which appears randomly, but is accepted as the inherited information because it proves to be useful. Among all small changes those which are favorable occur with sufficiently large probabilities. But it is improbable that randomly may occur the qualitatively more complex structure of organism, which nevertheless would prove to be more viable.

But how then do appear the actually observed jumps of evolution, which are accompanied by passages to more complex structures of organisms? The mechanism of super-cognition may act in this case too.

This mechanism is applicable in this case too because it is based on the usage of enormous, actually infinite, database of alternative scenarios. Actually something similar to the random changes and the subsequent selection occurs also in this case, but this happens not in one classical reality (which would be impossible), but with the sorting of all alternative scenarios (i.e., all possible chains of alternative classical realities relating to different moments of time). The "objects" that are sorted in this case are virtual.

10.2.2 *Miracles*

In the Quantum Concept of Consciousness it is assumed that in the state of unconscious the *super-information* (which cannot be obtained in the usual way) is extracted from the parallel worlds. This extremely valuable information can (in full or in part) arise in the conscious state. This phenomenon, subjectively perceived as super-intuition, was discussed in Sect. 10.2.1.

However, in the Quantum Concept of Consciousness and Quantum Concept of Life one additional assumption is accepted, according to which the information obtained from the parallel worlds can be used not only in the usual way (with the aid of the appropriate conscious actions), but also by an increase in the subjective probability (probability to subjectively perceive) for those parallel worlds, which are favorable for the life.[9]

If due to this ability such a reality is subjectively perceived which has objectively extremely low probability, then the sensation of the miracle, which occurs on the will of consciousness, is created. This may explain many actually observed "strange" events as well as the miracles, evidence about which can be considered reliable.

---

[9]This can occur also directly in the state of unconsciousness. The ability of the selection of favorable parallel realities (the ability possessed even by those living beings, that have no consciousnesses at all) is in fact the essence of the phenomenon of life.

10.2.2.1 *Miracles and science*

By an important example of the miracles, about which there are written evidence, appear biblical miracles, i.e., the miracles, described in the Bible. Biblical miracles are important because they form one of the supports of the Christian religion, i.e., of the ideology, by which the huge number of people adheres to.

True, the role of biblical miracles in the religion is frequently exaggerated. For example, Bible begins with the story how God created world in six days. It is assumed, apparently, that Christians must implicitly believe in this. But is this compulsory for the believers? From the other side, the atheists, attempting to refute the dogmas of Christian faith, frequently indicate that the creation of world during six days is the myth, which cannot be believed. But even if so, does this refute Christian religion as such?

Certainly, neither of these conclusions is correct. The story about creating the world in six days may be understood as a metaphor, and then for a believer it is not compulsory to believe in this story, understood literally; and its nonacceptance does not completely refute religion.

Similar status can be given to many other biblical miracles. They may be treated either as metaphors or exaggerations, aiming at illustrating various ideas (a kind of the parable).

However, some biblical miracles are important for the Christian religion (for example as proof of God's power), and miracle as such is an indispensable attribute of any religion, one of the important manifestations of the mystical aspect of religion. It is possible to say that without miracles there is no religion. Therefore it is important to give an estimation to the frequently meeting opinion that the miracle in principle contradicts science. Indeed science in our time uses the absolute trust of the enormous majority of people. What should they think of miracles and of religion?

It seems at first glance that the miracles are incompatible with the science by definition. Indeed, a miracle is just what cannot be explained, i.e. something for which there does exist no scientific explanation, that cannot be included in the scope of science. However, it occurs that this is erroneous for the phenomena, which are predicted in the Quantum Concept of Consciousness and Quantum Concept of Life and which we named *probabilistic miracles*.

The reason for this is very simple. A probabilistic miracle is an event, which does not contradict the laws of natural sciences, but probability of which, according to these laws, is extremely small. If improbable event occurs, then it subjectively looks as a miracle, if it occurs "in tight time", just when people badly need it or when some person, "miracle worker", passionately desires this event. But then a skeptic, who does not believe in miracles, may say that it is not a miracle, what happened, but coincidence.

The compatibility of probabilistic miracles with the laws of natural sciences becomes even more convincing because in quantum mechanics the events of probabilistic nature, i.e., random events, are not exceptions, but the rule, and the result of observation is random event even when the state of the observed system is precisely known.

Thus, a probabilistic miracle is what from the other side can be interpreted as a rare coincidence, which, nevertheless, does not contradict the laws of nature because these laws have the probabilistic nature.

The above reasoning shows that probabilistic miracles are possible, even from the point of view of the scientific views. But within the framework of the Quantum Concept of Consciousness and Quantum Concept of Life the probabilistic miracles not only are possible, but they must occur, moreover, they are the very essence of the phenomenon of life. Customary phrase "life is a miracle" acquires within the framework of QCL the status of precise assertion rather than metaphor.

10.2.2.2 *Biblical miracles*

Some miracles, which are described in Bible, indicate the power of God and confirm that God patronizes to the people. Such indications and the confirmation are important for Christianity; therefore the stories of this type miracles are aimed for the believers to perceive them not as metaphors or parables, but as stories about the real events. But are such events possible?

Let us consider only one example, the episode of the exodus (escaping Jews from Egypt), since many others may be analyzed in the same way.

Moses conducts the escaping Jews, with the army of Egyptians following them on the heels, but Red sea intercepts to the fugitives. However, Moses raises prayer to the Lord about rescuing his people, and God accomplishes the miracle: the dry corridor through the sea appears letting Jews to cross it, but then water returns again to stop the army of Egyptians. Here is the corresponding text of Bible:

> "Then Moses stretched out his hand over the sea; and the Lord swept the sea back by a strong east wind all night and turned the sea into dry land, so the waters were divided. The sons of Israel went through the

midst of the sea on the dry land, and the waters were like a wall to them
on their right hand and on their left." (Exodus 14:21-22)

It seems at first glance that drying sea in this episode is a real miracle in the sense that it has nothing to do with the natural way of being
and therefore incompatible with the scientific laws. However, the thorough
investigation of the place, where the way of Jews might lay, showed that
sometimes there blow the high wind, which blow out water, baring the bottom of sea. This phenomenon can be characterized by words "turning the
sea into dry land".

The specialists, who revealed this strange phenomenon, declared that
the biblical story describes not a miracle, but the real event, which occurred
completely naturally and it is quite agree with the laws of nature. Thus,
there was no miracle whatever?

Certainly, this conclusion is incorrect. The wind, which bares the bottom of the sea, rises in this locality very rarely. Why this wind did begin
precisely when Jews did approach this place? Why it did end precisely
when Jews had passed along the bared bottom and Egyptians came? Of
course this looks like the miracle, which occurred in response to the prayer
Of Moses. But the miracle is not in the fact that the sea made room, but
in the fact that this completely natural event began at the necessary time
to make it possible for Jews to cross the see, and it ended in time in order
not to give for Egyptians to do the same.

Thus, we have an example of probabilistic miracles. Skeptics will say
that Jews were lucky to meet the rare coincidence (of the time of the Moses'
prayer with the time of the wind). But from the point of view of the
Quantum Concept of Consciousness and Quantum Concept of Life nothing
else than a probabilistic miracle occurred: the anxiety of Moses brought the
method to overcome the obstacle, increased probability that Jews "turned
out to be" footnote let us recall that word they proved to be it relates not
to objective reality (which contains all possible versions of the course of
events), but to that of parallel realities, which survives subjectively. in just
that of the parallel worlds (in that of alternative realities), in which the
wind necessary for their rescuing rose and ceased precisely at the moments
which were necessary for their rescuing.

If we actually accept this picture of what happened, then there does not
exist, and cannot exist, any way to prove that it was only a simple coincidence and not miracle that occurred. At the same time it cannot be proved
that it is precisely miracle that occurred (indeed, random coincidences also
happen). It is not difficult to see that the associated circumstances make

subjectively (but not objectively!!!) much more convincing the version, that in this case a probabilistic miracle occurred, that was caused by the passionate desire Of Moses at the moment of its prayer.

10.2.2.3 *Good weather etc.*

Let us consider now more prosaic, each-day matters that we see around ourselves in the usual life. Are there in this life miracles or at least such strange, inexplicable things as super-intuition? The absolute majority of people will say that they of course met nothing similar. But could not it be that we simply insufficiently attentively do look all around?

I know someone who in summer almost never takes an umbrella with himself. He does not need an umbrella because rain can begin only when this man is on the work, either in the transport, or visited into the store after the purchases. When he leaves to the street, rain ends. Explaining how this happens, he told me that when leaving home, he does not think about an umbrella, but if his hand is reaching after the umbrella without his will, then he takes it, and this means that he will be caught in the rain. It is important in this concretizing that the decision to take the umbrella along is accepted unconsciously, which is the sign of resorting to super-intuition.

When this person is going to visit his country house, I know that the weather will certainly be remarkable, even if the forecast, which gives the weather bureau, is not very good. It is deep autumn now, and the weather is autumnal. But recently my familiar went to countryside for one day. This entire day the sun shone, although before this no sunny days occurred more that for a month, and already on the next day after his trip it was again cloudy.[10]

But how all this can be explained? If these are random coincidences, then why they do occur with this regularity? It is understandable that the probability of the large number of this type coincidences is very small. Rather we meet here an example of special abilities of consciousness of the type of super-intuition or probabilistic miracles. True, in this case these abilities are manifested not as vividly as those which are usually called miracles. Besides, the man I tell about reveals capability for clairvoyance also on other occasions, which do not refer to weather.

---

[10]I think that at this moment many readers will think that I could simply not notice or not memorize those sunny days that happened in the preceding period by which were not connected with this special case as the trip of my familiar to countryside. However, this is not so. Just before my friend's trip I happened to hear, as a comment in the weather forecast, that the sun did not appear already during 42 days.

The abilities of my familiar are by no means unique, and I shall present one more example. I had a friend, in many respects surprising person, who in the literal sense knew how "to govern weather". Once he happened to travel on the motor ship along the river, and the weather was permanently rainy or cloudy. However, each time, when the ship moored for the passengers to be able to take a walk on the shore, the clouds were scattered and the sun shone.

This my friend (now deceased), Boris Viktorovich Vedmin (his family nave, Vedmin, is Russian for "a son of witch"), was in many respects unusual man. He lived in the ancient town Sergiev Posad near Moscow. The town is built around the known Troitse-Sergiyeva Lavra (Holy Trinity-St. Sergius Laura), the monastery, which played enormous role in the history of Russia. The excellent white-stone Troitsky cathedral stands in the middle of Laura, with the most, perhaps, known in Russia icon of the holy trio of Andrey Rublev's brush. All this beauty was captured by Boris Vedmin in surprising photographs. He worked as a building engineer, although the artistic photograph was the aim of all his life. He told that he did not become professional photographer on purpose, in order to be completely free in photography.

In the last years of his life, Boris Vedmin headed the building of the excellent hospital complex in Sergiev Posad, and the title of the honorable citizen of this town was appropriated to him. In this reward was also taken into account another merit of Vedmin before the city: his photographs imprinted the unusual beauty Of Sergiev Posad and, first of all, Laura.

Boris Vedmin had the beautiful tradition: almost each year he, together with the wife, attended the Solovetsky monastery, located on the island in the north of Russia. There Boris Vedmin also made numerous surprising photographs, which imprinted the beauty of island and monastery. Solovetsky monastery is one more sacred point in Russia, not less known, than Troitse-Sergiyeva Lavra, although the history of this monastery is tragic: in it was organized one of the first concentration camps, in which, in particular, Pavel Florensky, the mathematician, philosopher and priest, was a sentenced prisoner and was shot.

Several years ago the widow of Boris Vedmin, Alla, presented me a collective volume about the history and the present of Solovetsky island. The volume included also articles about Boris Vedmin who was well-known in the island. A small episode characterizing his visits to the island struck me. Two women meet in the street, and one of them says: "Do you know that Boris Viktorovich arrived?" The second woman answers: "This is fine! This means that now for a long time a good weather will be established".

10.3   Discussion

Let us briefly discuss the contradiction that seem to exist between mystical features of consciousness and the laws found by natural sciences (this issue will be discussed in more detail in Conclusion).

One more comment will be about quantum computers. We shall argue that a model of life including its special quantum features (as they are assumed by our concept) can be constructed with the help of quantum computers (this has been discussed also in Sect. 5.2.3).

10.3.1   *Consciousness and the laws of natural sciences*

Thus, it follows from the Quantum Concept of Consciousness (QCC) that the consciousness must possess the special possibilities (super-intuition and probabilistic miracles), which at first glance seem impossible, because they seem to contradict the laws of nature. In principle the contradiction could exist, because in the course of constructing QCC two arbitrary assumptions were made, so that QCC does in fact go beyond the framework of quantum mechanics. However, there is no contradiction in reality. On the contrary, the strange abilities of consciousness predicted by QCC improve the status of quantum mechanics making natural and even necessary the strange, counter-intuitive features characteristic of the quantum mechanics itself.

Besides this, the strange features of consciousness are in fact confirmed. Long ago these features were already noticed by people and are being studied in such spheres of knowledge as religions, eastern philosophies and mystical schools. In the scope of these doctrines the above-mentioned features of consciousness are known as its mystical issues.

But the most important is that the detailed analysis reveals manifestations of these features in our usual life. We meet these manifestations, perhaps, not very frequently, but also not as rarely as it seems at first glance. However, people, whose world views are based on science, when being encountered with mystical phenomena, explain them as simple coincidents.

True, there are reasons for this view. The laws of nature, which are acknowledged as science, always bear probabilistic nature.[11] In view of this, the events looking as mystical always may be interpreted as random

---

[11]In classical physics this is connected with the fact that the state of any real system is never known with the absolute accuracy, while quantum mechanics showed that the probabilistic nature of observations proved to be a fundamental property of nature.

coincidences, and it is impossible to prove with the absolute certainty that they are actually mystical events rather than coincidences.

As for the impossibility to make the specific conclusion about reality of mystical phenomena, this impossibility is extremely interesting in its own right. It shows that the natural sciences are connected with the sphere of spirit very softly, without the clearly determined boundary. Where these two spheres come in contact with each other, the region of uncertainty lies, which is common for both spheres.

Each of the two spheres can quite satisfactorily be developed independently of another, and only deep conceptual analysis of each of these spheres detects in each of them logical defects, disappearing after their unification. Materialism and idealism become the relative concepts, each of them being applicable in its own area. The area where both of these concepts are applicable is the sphere of mystical phenomena. Anyway, their contradistinction is in our time counter-productive.

10.3.2   *Quantum computer: model for consciousness (for physicists)*

Quantum computer may be used for modeling the 'quantum consciousness' as the latter is assumed in Quantum Concept of Consciousness (QCC). Indeed, according to Everett's interpretation of quantum mechanics, all classical alternatives evolve parallely and independently from each other. It is assumed in QCC (generalizing Everett's interpretation) that 'consciousness' is nothing else than this independence (separating the alternatives from each other). The 'super-consciousness' is, vice versa, unity of all the alternatives as components of a superposition. Both the separation (independence) of the 'alternatives' from each other and their unity in the superposition may be illustrated in a quantum computer as a model. This could experimentally demonstrate at least the fundamental possibility that such 'quantum consciousness' may indeed exist.

This structure may be realized in a quantum computer in the following way. The quantum states evolving in a quantum computer are superpositions with a large number of components. Each superposition component carries some classical information (e.g., a binary number) and the evolution of the entire superposition ensures quantum parallelism, i.e., the simultaneous transformation of all these variants of classical information. In the model of quantum consciousness, individual superposition components can model the alternatives into which the consciousness divides the quantum state. The information contained in each component is a model of an 'alternative classical reality', i.e., the alternative state of a living creature and its environment.

The problem in creating the model of this type is 1) to formulate a criterion of what will be called survival, and 2) to select the evolution law such that the evolution of every alternative (superposition component) be predictable, and survival in this evolution be possible (although not guaranteed).

Of course, the task of constructing such a model is by no means simple, but it is basically solvable using a quantum computer. It is well-known that 'big' quantum computers, which promise extraordinary new capabilities, have not been realized. However, this applies only to quantum computers with the number of cells of the order of a thousand or more. As for quantum computers with the number of cells around ten, they have already been realized. Evidently, the number of cells attained will increase further, though maybe slowly. It is conceivable that even with these 'low-power' quantum computers, which will be constructed in the relatively near future, it will be possible to realize the model of 'quantum consciousness'.

Chapter 11

Conclusion: Science, philosophy and religion meet together in theory of consciousness

The Quantum Concept of Consciousness (QCC), presented in this book, is based upon the conceptual structure of quantum mechanics, but its conclusions relate to a completely different sphere, the sphere of psychology or, more generally, to the sphere of the spiritual life of human beings.[1] Therefore, it is possible to comment on contents of the book from two different points of view: from the point of view of quantum mechanics and from the point of view of psychology. For this reason some sections of the present concluding chapter will be oriented rather to the readers that are physicists (although they will be intelligible by all) and, the other to all readers.

But, first of all, let us examine a question about what is the practical value of the developed concept, QCC.

11.1   Why QCC is necessary, or how to learn to believe?

At first glance QCC is a purely theoretical development, which has no practical value. Although the special abilities of consciousness are very important from a practical point of view, but the concrete methods of their application are familiar and are developed in the the framework of religion, eastern philosophies, various spiritual practices and even such ancient beliefs as shamanism.[2] What new, in comparison with the recommendations of these teachings, can give the support, on the basis of quantum mechanics, of the very hypothesis of mystical abilities of consciousness?

---

[1] When presented with detailed analysis of the quantum-mechanical aspect, as in Part 2, this approach is called Extended Everett's Concept (EEC).

[2] According to a thesaurus, shaman is a priest or priestess who uses magic for the purpose of curing the sick, divining the hidden, and controlling events.

In order to answer this question, let us note that all mentioned spiritual schools and practices include the requirement of faith as a necessary and extremely important element. The concrete formulation of faith can vary. This can be the faith in the one God (Soli Deo, as in the monotheistic religions), or faith in many gods (as in various polytheistic religions and in paganism), or faith in the Truth and in Way (as in Buddhism and the doctrines close to it), and so on. But the element of faith is always required.

As a rule, in the enumerated studies and the practices, some dogmas are formulated, the faith in which is required. The least dogmatic is the approach to a question of faith, which is practiced by *Buddhists*. Teacher tells his scholar: believe nothing that you will hear from me until you are convinced of this from your own experience.

But how can the scholar make certain about what the teacher tells him? In the process of his practice, which consists in the work with his own consciousness (mind). This is sophisticated process, and the role of teacher in this process is special, because he cannot unambiguously formulate the final goal, which the scholar must approach. The teacher helps him only with the aid of the analogies and metaphors, as well as by controlling his purely physical actions, which contribute to reaching the necessary state.

Significant time is needed for the achievement the purpose in this process of training or rather self-training. But finally, in case of a sufficient patience and zeal, the scholar sees in his own practice that those states of the psyche, about which the teacher speaks, are actually feasible. Gradually he masters the methods, which make it possible to approach these states promptly.

Then it is reached what the teacher spoke about from the very beginning: the scholar believes in what the teacher speaks about, because he himself proved that this was true. Of course his faith is extended now also on those statements of the teacher, which he did not yet verify. The most important is that the scholar masters now the lesson: instead of the blind faith in these or other dogmas of *Buddhism* he can verify them in his spiritual practice.

The way, which leads to the faith, is in this case complicated and long, but it gives very solid grounds for the faith. This is more difficult, but also it is more reliable than to attempt from the very beginning to blindly believe in what the teacher speaks. In the majority of religions the persistent requirement is practiced to blindly believe in the dogmas of this religion. This easily achieves the goal in case of the people who easily yield to suggestion, but for thinking people this makes the faith less solid. Clear example is Leo Tolstoy, for whom the doubts about the faith and overcoming of these doubts were difficult vital task.

Thus, by one or another method the adepts of various spiritual schools reach faith. What they believe in, may be called in various ways. But the essence is the same in all cases: this is always the faith in the mystical component of the corresponding doctrine. In the wider formulation this can be named the faith in the mystical features of consciousness. In reality this is the same, because the mystical component of consciousness (including unconscious) cannot be limited to body. It includes the entire world. In this sense each person is the entire world, microcosm. In this sense each person is God.

Reservations here are, true, necessary. Human Being is God only he/she rejects the selfish position and identifies himself/herself with all living. God is the good, and man is God, if he/she represents the good. If the person appropriates the right to arbitrarily determine what is the good (to determine it from the selfish either from the group positions or even from the positions only of people, ignoring remaining living nature), he/she accomplishes the first-born sin (see Sect. 9.1.6). But if he/she for the whole life is learning to define the good, taking into account the interests of the entire living, then this may be called the tendency of man toward the God, as the unquestioning obedience to God.[3] And then it is possible to say that human being is God.

But why faith is so important, practically necessary? Because it is the necessary condition for the mystical abilities of consciousness to be effectively realized, practically working. Both super-intuition and ability to make probabilistic miracles are inherent for any person. But if he/she believes to actually possess these abilities, they become more accessible.

This is well illustrated by that fact, which is now widely believed. Each child is capable of the mysticism, but the majority of parents destroy in it these abilities depriving them the faith in the mystical phenomena they perceive.

And here we can return to the question, what is the practical value of the Quantum Concept of Consciousness. Of course, it is in that this concept liquidates the inconsistency between science and the extra-scientific sphere of knowledge, between the material and the spirit.

Many people consider this inconsistency to be an unquestionable fact. Can such people believe that their own consciousness (mind) does possess mystical abilities? Either they cannot believe this, or they can, but not entirely. But then they cannot use these possibilities in full measure.

---

[3]Of course, one cannot learn to this finally. According to the Leo Tolstoy's diary, he suffered from not being able to love everyone and every animal he met.

QCC makes it possible to understand that there is no contradiction between the science and the mysticism.[4] This makes it possible for them to believe (in God, or in Truth, in Way of Buddhism, and so on) and offers the enormous possibilities, hidden in human beings, the possibilities, which make one truly free, possibilities, without which he/she is only a slave of the external circumstances.

By the way, for the scientists, and even for other people, this gives the possibility to make use of very effective methods of creative work.

11.2   Science and mystics

The main task of this book is to show that mystical features of consciousness (mind) do not contradict to natural sciences. In previous chapters we shall consider this issue from various points of view. Now we shall briefly present some of these arguments, illustrating them in as simple way as possible.

11.2.1   *Why physicists do not believe in the miracles*

Our world is quantum, independently of whether we want this or not. Reality, which rules in this world, differs from that concept of reality, which is accepted in classical physics. This *quantum reality* is presented by parallel (Everett's) worlds. Many physicists meet into the bayonets the interpretation of Everett and all the more its generalizations of the type of that, which is developed in this book. Why?

Because the intuition of these physicists, their work experience in physics, speak that nothing similar to parallel worlds has never been observed, and they believe it cannot be observed. And they are right: nothing similar can be observed with the help of the usual method, i.e., by fixing the events with the aid of the instruments and then by perceiving the readings of these instruments.

However, it is erroneous that it is not possible to perceive such phenomena by consciousness. Why? Well, simply because we do not know what consciousness is.

Most of physicists reject any specific consequences, connected with the interpretation of Everett. They think that the Everett's interpretation differs by wording, but it does not differ by formulas and prescriptions

---

[4]On the contrary, science needs mysticism for its logical completion, we will speak about this in Sect. 11.2.

for calculations; therefore it cannot be distinguished from the Copenhagen interpretation with the aid of experiments.

Roger Penrose, the famous mathematician, wrote three books about human mind [Penrose (1991, 1994, 2004)]. He believes that consciousness must be somehow connected with quantum mechanics. What about the Everett's interpretation, he argues that we cannot say even, what consequences this interpretation can produce, until we know what consciousness is. We must, according to Penrose, construct theory of consciousness, and only after this we may return to the estimation of the Everett's interpretation.

What is made in our book (and in the previous works of the author), is the third way: the construction of the theory of consciousness (at least its basic condition) with the support on quantum mechanics and then the derivation of consequences of the resulting theory of consciousness together with the Everett's interpretation.

The theory of consciousness cannot be literally derived from quantum mechanics. However, the interpretation of Everett hints, what the theory of consciousness must be. The conclusion, based on thus obtained theory of consciousness, lies in the fact that with the aid of the consciousness we can directly observe the features that distinguish quantum reality from the classical one. With the aid of the consciousness (but in the process of going over to unconscious and backward) we can observe the phenomena, connected with quantum reality, the phenomena, which are possible because in our world quantum reality rules.

The conclusions, which follow hence, are cardinal and very interesting. They explain numerous strange phenomena, which are described in the mystical teachings, including religion and oriental philosophies. These phenomena are then treated as the manifestations of quantum reality in our world.

Particularly such very strange type of the phenomena as probabilistic miracles is connected with quantum reality. A probabilistic miracle is an improbable event, which nevertheless actually happened in "suitable" time, when there was urgent need in it. Are probabilistic miracles possible? Are not they contradict to the lows of science?

It turns out, that 1) it is impossible within the framework of the scientific methodology to prove that such phenomena cannot occur, and 2) if such a phenomenon occurs, then it cannot be said, is this phenomenon a miracle or a simple random event. Random events are allowed by quantum mechanics and compose one of the most strange aspects of this theory, which confused physicists for a very long time, but which is experimentally confirmed.

The probabilistic nature of the predictions of quantum mechanics is confirmed by many experiments and in fact by the enormous number of practical (technical) application of quantum mechanics. Therefore, our quantum world allows random events. Certain part of these random events may be probabilistic miracles. We can subjectively ascribe them to the probabilistic miracles.

The interpretation of such events as probabilistic miracles can sometimes be subjectively convincing. This may be convincing in view of associated circumstances, but we in principle cannot objectively prove that these are actually probabilistic miracles, caused by consciousness, and not simple random events. From the other side, we in principle cannot prove, that these are random coincidences and not probabilistic miracles.

The Hegelian triad (thesis — antithesis — synthesis) appears in case of each event of this type:

Thesis: Believers will say that miracles sometimes occur.

Antithesis : Physicists will say that miracles do not occur, but what happens are rare random coincidences.

Synthesis: Probabilistic miracles principally cannot be distinguished from the random coincidences.

The subjective sensation that a miracle occurred cannot be converted into the proof of the fact that this is actually a miracle. From the other side, physicists also cannot ever prove that single events or even finite series of events resembling miracles are actually only random coincidences, that their realization is not connected with consciousness.

Within the framework of the strict scientific methodology, physics can explain only the simplest phenomena, characteristic for the simplest physical objects. The complicated systems, whose manifestations are always individual, in principle cannot be explained by physics within the framework of the strict scientific methodology. However, going beyond the framework of this methodology leads to the conclusion that the miracles are possible.

11.2.2 *'Soft' embedding of life into the objective world*

The scientific laws of classical physics are deterministic. The systems satisfying these laws are similar to the mechanical devices (machines) consisting of hard details precisely fitted to each other (imagine gear engagement in which toothed wheels are caught with each other in a quite determined way).

Contrary to this, everything is soft in constructions used by life (imagine joint of limbs with the body and generally soft junction of bones and other parts of a body of any living being). Everything is soft also in the laws governing life: the future of the living system is never predetermined completely by the given present state.

Whether then life may exist in the world governed by deterministic scientific laws? No, this would be impossible if the laws of science were deterministic. However, the real scientific laws are not classical, They are quantum, and quantum world is in a sense not determined. But in what sense?

The laws of *evolution* of quantum system (Schrödinger equation for example) are deterministic. However, the results of *observation* (measurement) of any quantum system are not deterministic. Observation of the quantum world is governed by stochastic, of probabilistic laws.

Thus, the world we are living in is deterministic, but it looks (in observation) stochastic. This world is objectively hard, but subjectively it looks soft. Thus, quantum character of the world makes possible soft embedding of life in it, soft junction of life with the objective world. Our book is about this embedding, or junction between life and the objective world as its environment.

It is important for the way of junction that life is essentially subjective. More precise formulation for this is as follows. Quantum world is the set of parallely existing alternative (Everett's) classical worlds. Living beings live in these parallel worlds which are separated from each other. However the living beings subjectively perceive each of these parallel worlds independently of the others.

11.2.3 *Quantum paradoxes are compensated by mystical features of consciousness*

Soft junction of the laws of evolution of matter with the laws of life are especially interesting if we consider the central part of both spheres: quantum mechanics governing evolution of matter and consciousness as the manifestation of the highest form of life, humans. Both quantum mechanics and consciousness have some "defects" when being considered from the traditional viewpoints, but these defects disappear in junction of the two spheres, because the"defects" of one of these spheres compensate the "defects" of the other sphere.

Quantum mechanics has two specific features that radically differ the quantum phenomena from what we know in everyday experience and what is accepted in classical physics. These two special features of quantum mechanics are

- stochastic character of the observations in quantum mechanics and
- coexistence of macroscopically distinct classical states of the world.

These two features of quantum mechanics are usually treated as its logical defects (*paradoxes*) that have to be overcome in some way or another. This may be expressed as two cracks in the otherwise smooth body of quantum mechanics (Fig 11.1 left).

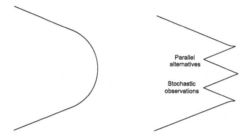

Fig. 11.1　Quantum mechanics (right) contrary to classical physics (left) has two features treated as logical defects: coexisting of classical realities and stochasticity of observations.

Mystical features of consciousness assume two phenomena that seem strange and quite impossible from the scientific viewpoint (and in this sense may be considered to be "defects" in the description of consciousness):

- miracles, i.e., the events which are caused by power of consciousness (mind) although they can happen in the natural way only with negligible probability, and
- super-intuition, or direct vision of truth (not based on any information available in observations).

These features may be symbolically presented as the two barbs at the otherwise smooth body of the spiritual knowledge (Fig. 11.2)

These two conceptual structures, illustrated geometrically, fit each other perfectly so that the smooth character of the body restores after their junction, or unification (Fig. 11.3).

The first of the mentioned features of quantum mechanics is common for all variants (interpretations) of quantum mechanics and makes possible

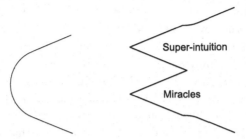

Fig. 11.2 Spiritual knowledge (right) has two mystical features: super-intuition and probabilistic miracles, that are not recognized in the "classical" understanding of the spiritual abilities (left)

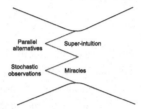

Fig. 11.3 Quantum mechanics and spiritual knowledge united into a closed logical structure: the "defects" of quantum mechanics explain the mystical features of consciousness

probabilistic miracles. The second one is accepted in the Many-World (Everett's) interpretation and makes possible super-intuition (direct vision of truth).

11.2.4 *Buddhism*

One of the spiritual schools, in which its philosophical aspect is deeply analyzed, is *Buddhism*. Some of the physicists and some of the Buddhists both recognize that there exist explicit analogies between the statements of quantum mechanics and the views of Buddhists on the features of material world, human consciousness and methods of the cognition, which are available for people. These analogies are traced in detail in the book of Alan Wallace [Wallace (2007)], who is Buddhist and who has been educated as a physicist. Now he is President of the Santa Barbara Institute for Consciousness Studies. Here are some quotations from this very interesting book.

Describing the conversation of Dalai Lama with the known quantum physicist Anton Zeilinger, Wallace writes:

"In the dialogues between Anton Zeilinger and the Dalai Lama, both were struck by this extraordinary convergence between quantum physics and Buddhism. As Piet Hut, another physicist at the 1997 meeting, commented, this could be a mere coincidence, but only if the physical world and the mental world are absolutely different without any possibility of transformation. If the themes of relativity and transformation are connected, then their convergence is not accidental. This could imply, he continued, that we are moving from a science of objectivity to a science of intersubjectivity, in which the next relativity theory will include a relativity between the object and the subject, between the physical and the mental."
([Wallace (2007)], pages 95-96)

He writes further:

"The intersubjective nature of the natural world does not imply solipsism in either physics or Buddhism. Laws regulating the interactions among physical phenomena, among mental phenomena, and between physical and mental phenomena can be discovered that are invariant across multiple cognitive frames of reference."
([Wallace (2007)], page 97)

Wallace comments also the concept of consciousness suggested by the present author:

His own [Mensky's] theory, which he calls the Extended Everetts Concept, makes new predictions not found in usual quantum mechanics, but they are for features of consciousness rather than for the results of physical experiments. Therefore, according to Mensky, his theory can be tested using methods found in Buddhism for observing human consciousness.
([Wallace (2007)], page 102)

11.3   Science and religion are compatible

Religion was the mainstream ideology during many centuries (if not millenniums), but in 20th century scientific viewpoints became in fact new religion that often pretends to be the only possible. Many people think that lack of belief was the reason of Bacchanalia of violence characteristic of the 20th century. The cult of science together with the opinion that religion and science contradict to each other, might play the essential role in this tragedy. It is thus very important to make clear the latter question.

11.3.1 *Basic aspects of various confessions*

Mystical features are necessary components of any religion, and this hints that it is one of the most important part of all the religious doctrines. The mystical features of these doctrines are important because they concern with the questions of ontology, to the understanding of reality accepted by one or another religion. And this is the reason why these features are common for all beliefs.

However, various confessions differ by many other features, first of all by their dogmas, rituals and other details of their forms. These features may be different because they have no direct correspondence to the concept of reality (although they may be essential for making the dogmas of a church transparent and thus easily acceptable for people).

This common idea of reality, strange for the modern science but evident for all religions, is the most reliable basis for contacts between various confessions. This is why ecumenism can become real future for many of them. This is very important in the contemporary state of society, when amusing development of science and technology led not only to great success in understanding material world but also to dangerous problems for human beings and actually for life on Earth.

11.3.2 *Science and religion need each other*

In the 21st Century the society sharply needs restoring of the unity of knowledge and overcoming the precipice between "materialist" science and "idealistic" extra-scientific forms of knowledge, first of all — religion. This proves to be possible in view of the fact that the complex of the concepts, connected with the consciousness, is general for the science and religion. Moreover, quantum mechanics, which is the highest achievement of natural sciences, becomes conceptually closed only after the direct inclusion in it the "idealistic" concept of consciousness and explicit considering subjective aspect of physical experiments.

The role of consciousness may be adequately taken into account within the framework of Everett's Many-World interpretation of quantum mechanics. Concept of consciousness appearing in this case is so deep that explains the uncommon abilities of consciousness, which are manifested, in particular, as "probabilistic miracles". Thus, mystical aspects characteristic of any religion not only are compatible with natural sciences, but natural sciences (first of all their central part, quantum mechanics) is logically defective without inclusion of the concept of consciousness with its mystical features.

11.4   Philosophical viewpoint

Physicists do not usually recognize that the character of their work may depend on their philosophical position. The majority of physicists looks at the philosophy condescendingly, considering it not as science but the skill of the manipulation by words. However, the interrelation of physics and philosophy become significant for physicists and are actively discussed at the key moments of the development of physics.

Two aspects of these interrelations are then important. First, physicists discover that their specific philosophical views (or at least the methodological principles accepted by them) nevertheless influence their work (first of all this concerns the methods of interpreting experimental results. Second, the qualitatively new achievements of physics change the philosophical position occupied by most of physicists. These processes are tightly connected with a change in the methodology, which turns out to be necessary in connection with the new achievements. All this together is the passage to a new paradigm in the science.

The period of creating quantum mechanics (the first third of 20th century), which coincided with the period of creating special and general theory of relativity, was such critical epoch in physics. These enormous developments in physics, especially quantum mechanics, overturned the world view of physicists, particularly forcing them to abandon too limited understanding of materialism.

Apparently, physics experiences now a similar period of an active change in the paradigm, and particularly in philosophy of physics. The change is necessary because of realizing the close connection of quantum theory with the phenomenon of consciousness.

11.4.1   *Wigner*

In the seminal paper [Wigner (1961)] the prominent physicist E.P.Winer wrote:

"Until not many years ago, the "existence" of a mind or soul would have been passionately denied by most physical scientists. The brilliant successes of mechanistic and, more generally, macroscopic physics and of chemistry overshadowed the obvious fact that thoughts, desires, and emotions are not made of matter, and it was nearly universally accepted among physical scientists that there is nothing besides matter. The epitome of this belief was the conviction that, if we knew the positions and velocities of all atoms at one instant of time, we could compute the fate of the universe for all future. Even today, there are adherents to this view though fewer among the physicists than - ironically enough - among biochemists."

The final conclusions made by Wigner were not so close to the modern view on consciousness. Yet the paper of Wigner was very important because he was bold enough to refute traditional materialistic dogmas. He wrote:

"The principal argument against materialism is not that illustrated in the last two sections: that it is incompatible with quantum theory. The principal argument is that thought processes and consciousness are the primary concepts, that our knowledge of the external world is the content of our consciousness and that the consciousness, therefore, cannot be denied. On the contrary, logically, the external world could be denied — though it is not very practical to do so. In the words of Niels Bohr, "The word consciousness, applied to ourselves as well as to others, is indispensable when dealing with the human situation." In view of all this, one may well wonder how materialism, the doctrine that "life could be explained by sophisticated combinations of physical and chemical laws," could so long be accepted by the majority of scientists.

Philosophers do not need these illusions and show much more clarity on the subject. The same is true of most truly great natural scientists, at least in their years of maturity. It is now true of almost all physicists — possibly, but not surely, because of the lesson we learned from quantum mechanics. It is also possible that we learned that the principal problem is no longer the fight with the adversities of nature but the difficulty of understanding ourselves if we want to survive."

Wigner remarked that the experience of quantum mechanics is compatible even with solipsism, but not with materialism. I think that such statements had very strong influence on physicists working on the conceptual problems of quantum mechanics. Even if they had not immediately great response, they much widened radically the spectrum of possible ways of thinking on the problems of quantum mechanics.

It cannot be said that Wigner was the first, who expressed doubt about the applicability of the materialism (in that understanding of this term, which had been accepted among the physicists) for the interpretation of quantum mechanics. On the contrary, all, who seriously thought on the philosophical aspects of quantum mechanics, saw that the traditional approach of physicists must be radically changed (see for example the statement of Pauli cited below in Sect. 11.5.1 at page 233). But Wigner, apparently, went in this direction most boldly.

As to my own opinion, the word "materialism" may be well-applied even to the combined theory of material and living systems, but the meaning of this world must be very wide. In this case what was traditionally referred as idealism, in many cases may be treated as widely understood materialism. However, to be honestly, the concepts of idealism and materialism become relative and loss their importance.

11.4.2   *Objective and subjective*

The central point of the methodology of physics and more generally of natural sciences is the objective character of their laws. However, in the framework of quantum mechanics this became doubtful, because the conceptual problems (paradoxes) of quantum theory could not be removed without explicit inclusion of consciousness in the theory.

The formulation of quantum mechanics as the purely objectivistic science encountered formidable difficulties. The artificial character of this formulation is clearly seen, for example, in the following explanation given by Schrödinger:

> "Without being aware of it, we exclude the Subject of Cognizance from the domain of nature that we endeavour to understand. We step with our own person back into the part of an onlooker who does not belong to the world, which by this very process becomes an objective world."
> ([Schrödinger (1958)], page 38)

The difficulties with objective formulation was a hint that quantum theory, to be logically closed, should include not only objective, but also subjective elements. True, the probabilistic predictions of the behavior of quantum systems could be confirmed by repeated experiments and therefore were objective. But the results of each of the observation (measurement) could be fixed only subjectively, by the observer's consciousness.

This hint, treated in the framework of the *Everett's Many-Worlds interpretation* of quantum mechanics, became the starting point for formulating the basic points of theory of the conscious itself, as they are presented by Extended Everett's Concept, of Quantum Concept of Consciousness, suggested by the present author.

For estimating this theory we need new methodology including both objective and subjective elements. This implies the quite new situation when the new theory either should be treated as not including in the scope of various areas of physics or being included in physics but with the extended notion of physics supplied by the new (extended) methodology, recognizing subjective methods of investigation (observation of the observers own consciousness and transitions between conscious and unconscious states).

Let us remark in this connection that there is one more area of physics which also needs extended methodology. This is quantum cosmology. This branch of physics has been greatly developed in the last decades, because applying cosmic apparata for research of the cosmic background radiation (issued at the very early stage of the Universe's evolution).

It is sometimes said that quantum cosmology (treating early Universe as a quantum system) became an experimental science. This is because the character of Universe' behavior at the very early times after Big Bang (when Universe was quantum) following from the purely theoretical considerations, may now be confirmed by the characteristic features of the background radiation.

Thus obtained confirmation of theory by observations was great sensation some two decades ago. Yet the thorough analysis shows that the results of the observations may be considered to be confirmation of the theory only if the extended methodology is applied that admits as the criterion of truth not only series of repeated measurements but also single events (but having complicated structure). If not extend the methodology in this way, then the observation of the properties of Universe (for example the properties of the background radiation) cannot prove or disprove the laws of quantum cosmology. For reliable proof we need then a series of observations with many identical universes that is evidently impossible [Panov (2010)].

Nevertheless, the results of the observation of the background radiation and their agreement with theoretical predictions were so convincing that most people consider these observations to be the reliable evidence of the theory being valid (at least in general features). Yet, the fact of this agreement "being convincing" is not more than the subjective impression. In this case therefore a subjective convincing of something makes the physicists to depart from the commonly accepted methodology of the objectivistic science, to extend the methodology.

Thus, consideration of the features of background radiation as a confirmation of the laws of quantum cosmology is meaningless from the point of view of the standard physical methodology [Smolin (2009)], but rejecting these data as the evidences for quantum cosmology seems absurd and is not accepted by the most of the physicists working in this area. They prefer to extend the methodology, although they do not always clearly understand that they do this.

Therefore, the accepted methodology is not an inviolable law, its extension is possible when the subject of the theory is widen. In case of theory of matter and consciousness, or matter and life, the extension of the subject is much more radical than have ever took place in physics or other natural science. However, this extension is quite reasonable because its results turn out to agree with the whole experience of the mankind (although this is the experience in the spiritual sphere).

11.4.3 *Material and ideal*

"Consciousness", "subjective" are the concepts which evidently belong to the sphere of ideal. The numerous attempts to explain the phenomenon of consciousness as the result of the work of the brain, are in reality unfounded, if we have in mind the fundamental level of this phenomenon. Various rational thought processes, which occur against the background of consciousness, can be explained by the work of the brain as a material system performing a kind of computational operations. For fulfilling these operations the brain has, of course, the units for input and output of information as well as loops of feedback. However, this does not help to understand what is consciousness (otherwise it would be necessary to say that the computers possess consciousness too, which is intuitively incorrect).

From the other side, the analysis of the logical structure of quantum mechanics shows that it has in it the hint for the possible definition of consciousness, which will be ideally coordinated with this structure, makes it possible to simplify this structure, and moreover, gives interesting consequences. According to this definition, consciousness is the separation of alternatives. Then consciousness is something which leads from quantum reality (co-existence of parallel worlds) to the classical reality (subjective perception of only one of these worlds). This is just what may be expected as the ability of something that may be called consciousness: the passage from quantum reality to the classical perception.

The Quantum Concept of Consciousness (QCC) is based on this definition of consciousness. The detailed analysis of this concept shows that thus determined consciousness possesses mystical abilities (super-intuition and the ability to create probabilistic miracles). Thus, starting from the purely materialistic theory (quantum mechanics), we come to the ideal concept of consciousness and to the conclusion that the phenomenon of consciousness must possess mystical features, which are at first glance not at all compatible with the materialism.

It is obvious that the Quantum Concept of Consciousness (and in the more general Quantum Concept of Life) contains, in the close unity, as indivisible from each other, the elements, which traditionally are treated correspondingly as material and ideal. From the point of view of this concept, materialism and idealism lose previous meaning, they become relative.

The philosophical system, which is compatible with QCC and QCL, can be, if it is convenient, named materialism, but only if we radically broaden understanding of materialism. In any case, this is such materialism, which essentially includes subjective. Soft unification of quantum mechanics with mystical features of consciousness (sphere of science with spiritual sphere) makes uncertain, fuzzy, the boundary between material and ideal.

11.5 From quantum mechanics to consciousness

Let us comment on some points in history of ideas that made finally possible formulating Quantum Concept of Consciousness. We shall not follow this history in all its detail, but mention only the issues that illustrate, in some way or another, the status of our Concept.

11.5.1 *Pauli and Jung*

The basic idea of this book lies in the fact that the mysterious, mystical possibilities of our consciousness are explained by correctly understanding objective reality. The naive understanding of reality, which is based on the everyday experience and which is successfully adapted in classical physics, occurred to be erroneous. It is only quantum mechanics that gives correct understanding of what actually exists and what is only an illusion of our consciousness.

From the first years of existence of quantum mechanics this was manifested in the paradoxes. The paradoxes appeared in quantum mechanics, could not be removed and were invariably connected with measurements or observations. Therefore, the paradoxes appeared just when the physicists tried to describe how objectively existing reality may be reflected in the subjective perception of this reality by an observer.

Despite the fact that the corresponding questions were set already at the dawn of quantum mechanics, answers to them required many decades of experiments and several fundamentally new approaches, connected with the names of the outstanding physicists, beginning from Einstein. It is only in our days that the outlines of the concept appear, which makes it possible to answer these questions.

The key idea of this concept is that the phenomenon of consciousness can be explained only with the aid of the statements of quantum mechanics. The resulting theory indicates that the consciousness must possess mystical features, which are substantially connected with the unconscious. This theory, or Quantum Concept of Consciousness, outlined in this book, became possible only after the essential features of quantum reality having been formulated by Everett in the language of parallel worlds.

This makes really astonishing that one of the creators of quantum mechanics, Wolfgang Pauli, sufficiently accurately expressed the central idea of the quantum approach to theory of consciousness even before the appearance of the Everett's interpretation. Pauli arrived at this idea in the process of collaboration with the great psychologist Carl Gustav Jung.

Apparently, colleagues of Pauli considered his thoughts (concerning the direct connection of quantum mechanics with the phenomenon of consciousness) to be distrustful. It is possible that also himself considered these issues insufficiently investigated. In any case, Pauli never expressed himself on this question in scientific articles. His considerations concerning this question are known only from his letters to colleagues–physicists. Here are some of them.

In 1952, in Pauli's letter to Rosenfeld he wrote:

> "For the invisible reality, of which we have small pieces of evidence in both quantum physics and the psychology of the unconscious, a symbolic psychophysical unitary language must ultimately be adequate, and this is the far goal which I actually aspire. I am quite confident that the final objective is the same, independent of whether one starts from the psyche (ideas) or from physis (matter). Therefore, I consider the old distinction between materialism and idealism as obsolete."

[Letter by Pauli to Rosenfeld of April 1, 1952. Letter 1391 in [Meyenn (1996)], p. 593. Translated by Harald Atmanspacher and Hans Primas in [Atmanspacher and Primas (2006)]]

Still earlier he wrote to Pais [Letter by Pauli to Pais of August 17, 1950. Letter 1147 in [Meyenn (1996)], p. 152. Translated by Harald Atmanspacher and Hans Primas in [Atmanspacher and Primas (2006)]]:

"The general problem of the relation between psyche and physis, between inside and outside, can hardly be regarded as solved by the term 'psychophysical parallelism' advanced in the last century. Yet, perhaps, modern science has brought us closer to a more satisfying conception of this relationship, as it has established the notion of complementarity within physics. It would be most satisfactory if physis and psyche could be conceived as complementary aspects of the same reality."

These and close to them ideas of Pauli, expressed by him only very briefly and only in the letters, were very long practically unknown. Only in recent years, in connection with the increased interest in the quantum theory of consciousness, they became popular, so that articles and books (see for example [Atmanspacher and Primas (2006)] and [Enz (2009)]). The author of the present book learned about the statements of Pauli only in 2008, when several articles and a book on Quantum Concept of Consciousness were already published by him. The surprising agreement of this concept with the visionary thoughts of Pauli is additional confirmation for it.

11.5.2 *Penrose*

The well-known mathematician and physicist Roger Penrose was one of those, who in the recent decades made much in order to establish the connection between phenomenon of consciousness and quantum mechanics. However, he thought (and apparently thinks until now) that there are more questions than answers in this area. In the foreword to the book [Abbot, Davies, and Pati (2008)] he wrote:

"There is, indeed, a distinct possibility that the broadening of our picture of physical reality that may well be demanded by these considerations is something that will play a central role in any successful theory of the physics underlying the phenomenon of consciousness."

And further, characterizing this possibility more concretely, Penrose writes:

> "...are the special features of strongly quantum-mechanical systems in some way essential? If the latter, then how is the necessary isolation achieved, so that some modes of large-scale quantum coherence can be maintained without their being fatally corrupted by environmental decoherence? Does life in some way make use of the potentiality for vast quantum superpositions, as would be required for serious quantum computation?"

In this statement of Penrose possibility is examined that something similar to a quantum computer may exist in brain. At the same time he sees the difficulties, confronting this hypothesis. Further on Penrose considers possibility of more radical withdrawal from the standard quantum mechanics:

> Do we really need to move forward to radical new theories of physical reality, as I myself believe, before the more subtle issues of biology-most importantly conscious mentality-can be understood in physical terms? How relevant, indeed, is our present lack of understanding of physics at the quantum/classical boundary? Or is consciousness really "no big deal," as has sometimes been expressed?
>
> It would be too optimistic to expect to find definitive answers to all these questions, at our present state of knowledge, but there is much scope for healthy debate, and this book provides a profound and very representative measure of it.

Here Penrose does not discuss the possible role in the explanation of the phenomenon of consciousness, which may be played by the Everett's interpretation (as this is assumed in our Quantum Concept of Consciousness). In the book [Penrose (2004)] he concerns this question. His conclusion, however, is that, earlier than speaking about the usage of the Everett's interpretation, it is necessary to construct theory of consciousness.

This is the fundamental difference between the views, which are presented in our book, and by the point of view of Penrose. Instead of independently building theory of consciousness and examining the Everett's interpretation after this, the author of the present book proposed (in 2000) to extract the gists of theory of consciousness from the analysis of the Everett's interpretation. This way proved to be successful, because it led to the logically simple concept of consciousness, which explains the large number of phenomena, which are conventionally considered inexplicable from the scientific point of view.

11.5.3 Why Quantum Concept of Consciousness was successful

During last decades many attempts were undertaken to explain the phenomenon of life and in particular the phenomenon of consciousness on the basis of quantum mechanics. Studies in this direction were started already in Schrödinger [Schrödinger (1958)]. Schrödinger, in particular, for the first time indicated the important role of quantum mechanics in the fact that in the living systems can have stable discrete characteristics, which are necessary for the transmission of hereditary information. The contemporary survey of different approaches to this problem can be found, for example, in the book [Abbot, Davies, and Pati (2008)].

Examination and attempts to understand from the point of view of quantum mechanics such uncommon phenomenon as consciousness, is of course especially complex problem. One of the ideas, which they attempt to use for this purpose, is the assumption that some structures in the brain work as a sort of quantum computer.

In our view, numerous attempts to explain the phenomenon of consciousness, even with the attraction of quantum mechanics, gave much less impressive results, than the presented in this book Quantum Concept of Consciousness (QCC). This is due to the uncommon approach that has been used in the construction of this concept. This approach was completely not characteristic for physicists, but it proved to be successful for the solution of this problem.

Attempting to explain the phenomenon of consciousness, physicists go along the way, which is customary for them and seems the only possible. Explicitly or implicitly, they assume that consciousness is a function of the brain, which, therefore, can be explained, relying on the laws of motion of the matter, of which the brain consists. Maximum, that quantum mechanics can give with this approach, is an attempt to consider the brain not as usual (classical) computer, but as a quantum computer. [5] In this case purely physical problem appears — to explain, why decoherence does not appear, which unavoidably would destroy quantum coherence and convert quantum computer in a classical one. However, even if we are ignore this

---

[5]Versions of this type of construction may exist, but they do not differ qualitatively.

problem, it is not nevertheless obvious that the functioning of the brain as a quantum system explains consciousness.

During the construction of QCC (or Extended Everett's Concept, as otherwise this approach is called) reasoning was quite different. Not material objects, but functions were examined in the stage of constructing QCC. We analyzed those functions that have to exist in the theory in order to explain, first, that we know about our world from physics, and second, as we subjectively perceive this world. The basic function, which in this case must be explained, is consciousness, i.e., the appearance of the actually observed subjective picture of the world, on the assumption that objectively the world is such as it is described in physics, which is rested on the enormous experimental material.

Further, we can and must, it goes without saying, rest on the subjective idea about the consciousness, which each person from his daily experience has. But from the other side that function, which we want to name consciousness (and which consists of the transfer of the objectively existing world into the subjectively perceived one) must be described also in the terms, characteristic for physics. To accomplish this, the analysis of quantum physics was performed in order to find in it something that could play the role of this function. It turned out that this may be absolutely naturally and unambiguously found within the framework of the Many-World interpretation of quantum mechanics (Everett's interpretation).

The analysis of what can follow from this definition of consciousness, gives the unexpected result. It occurs that turning consciousness off (in the state of sleep, trance or meditation) or even simply its disconnection from a certain object makes it possible to go beyond the framework of that subjectively received and to obtain access to the entire objectively existing world. Then the super-intuition, or super-consciousness, appears, i.e. the information becomes accessible, which is principally inaccessible in the completely conscious state.

After thus determined function, named consciousness, is described, the process of constructing the theory could be finished. All necessary already exist in the theory of consciousness. But for comparing this theory with other approaches it is one may raise the question about what role the brain plays. And it occurs that the brain does not generate consciousness, but the brain is a tool of the consciousness. Besides usual functions of information processing, the brain forms the queries, which the super-intuition must answer, and interprets in the usual symbols and by the usual means the information, which appears as answers to these queries.

The starting point for the reasoning was the circumstance,[6] that 1) our world is objectively quantum and therefore it must be described in the terms of quantum reality (i.e., its state is described by a superposition of parallel, Everett's, worlds), but 2) subjectively only classical reality is perceived (the illusion appears, that there exists only one of the parallel worlds). The function, named consciousness, consists then of the separation of alternative (Everett's) worlds. The disconnection of this function (sinking into the unconscious) removes the separation, so that the access to the entire set of the parallel worlds appears. This gives the information, which is inaccessible in a single one of these worlds (that subjectively perceived). So we inevitably come to the conclusion that the consciousness (but more precisely, the complex of consciousness and unconscious) possesses mystical properties.

If we now look at the final construction (which arose with the analysis of functions), then it appears that the elements of two types coexist and are tightly interlaced in it — those, which are customary assumed to be material, and those, which are usually treated as ideal. It becomes understandable, why it would be difficult to come to this picture, if we assumed from the very beginning that there were only material objects (molecules, atoms, elementary particles) and everything else might be derived from the properties of these objects. Consciousness and life — this is what cannot be simply derived from the laws of material world (although existence of living systems does not, it goes without saying, contradict these laws). They must be postulated independently (in our scheme, as the corresponding functions).

But is not it nevertheless possible to formulate this concept, considering the laws of natural sciences fundamental? Yes, it is possible, but we have then to define life as a special phenomenon, which is presented by the subset of all possible scenarios of the evolution of matter. This subset is called sphere of life. It can be determined by the condition that all scenarios included in the subset (belonging to the sphere of life) satisfy the criteria of life, first of all — the criterion of survival.

This formulation (Quantum Concept of Life, or QCL) may be easily given after we have already arrived at it, having preliminarily constructed the theory (concept) of consciousness. Using this intermediate stage (QCC), we attain that the construction becomes quite plausible and the way leading to this construction practically inevitable. In this way only two arbitrary

---

[6]or, if you want, hypothesis, but which is confirmed by entire experience of quantum mechanics

assumptions have to be accepted, which form very rigid and simple logic scheme, but they give the huge amount of consequences as a result. These consequences are immediately identified with the well-known facts. True, these are facts from the sphere of spiritual knowledge, but in view of its millennial existence this sphere, in its main points, is not less reliable than the much younger (although supplied by precise methodology) field of natural sciences.

11.6   Second Quantum Revolution

The appearance of quantum mechanics at the beginning of the 20th century was the most great scientific revolution, which affected not only strictly physical laws, but world view of physicists, their philosophical positions. By the great merit of the creators quantum mechanics headed by Nils Bohr was that they succeeded in formulating the so-called *Copenhagen Interpretation* of quantum mechanics. It included the clear rules, which made it possible to make calculations of concrete quantum systems, laying aside the philosophical questions, which are concerned the philosophical comprehension of what stands after these calculations.

After obtaining the possibility to work, without worrying about the deep philosophical comprehension of this work, physicists nevertheless continued the attempts to improve the philosophical aspect of quantum mechanics. This led finally to the concept of *quantum reality*, i.e., to the understanding, that the idea about what "real existence" means, is in the quantum world something different than in the classical world. Finally the *Multi-World interpretation* of quantum mechanics (Everett's interpretation) was created, which suggests a convenient mathematical apparatus, which expresses quantum reality.

Several decades went by before the Everett's interpretation obtained the acknowledgment of the sufficiently large number of physicists. This occurred, in particular, because the new purely physical tasks, which were appeared in the field of quantum mechanics (first of all, the new kind of its applications, named quantum information theory), directly exploited the specific features of quantum reality. Quantum reality became necessary for the physicists practically, even at the engineering level. This required the comparatively simple formulation of what is understood under quantum reality, and many people understood that the Multi-World interpretation is such a formulation.

The mastery of Multi-World interpretation, experience of working with it, made it possible to understand how this interpretation is organic for describing the quantum world when it is necessary to keep your mind on the most striking specific features of this world, quantum reality.

In particular, this made it possible to more deeply understand the long ago confronting question about the role of the observer's consciousness in quantum mechanics. After this the question about what is consciousness as such, has been raised in a new way. And then completely unexpectedly the new possibility (actually even need) was opened - to directly connect quantum mechanics with the *theory of consciousness*. Moreover, it turned out that the direct connection appears between the laws of material world and the laws, long ago formulated in the studies about the spiritual sphere. Any reason disappears for the confronting between matter and spirit, materialism and idealism. On the contrary, it becomes clear that these two spheres of human knowledge require each other not only on the culturological level, but also in the gnosiological aspect.

As a result, the new understanding not only of consciousness and spiritual sphere of human, but also of the phenomenon of life is gradually coming.

Work on the mastery of these new possibilities, new directions for research, by no means finished. It only begins and undoubtedly requires great efforts of the specialists of various profiles. However, it cannot be overestimated that the very posing of the questions in the plan of the unification of material and spiritual, became now completely real. Besides, the powerful conceptual apparatus, developed in quantum mechanics can be used in this work.

Summing up the above said, one may with the complete right to conclude that the work on the improvement of the interpretation of quantum mechanics lifted in our time to the new level. Some fundamental questions, which appeared already in the period of the creation of quantum mechanics, remained not solved almost entire century. But now they gradually obtain their solutions, leading in this case to the enormous shift in the world view - to the direct unification of the material and spiritual.

There is no doubt that we become the witnesses of new scientific revolution. It can be estimated as the completion of that scientific revolution, which began in the period of the creation of quantum mechanics. Instead of this the new stage of quantum mechanics with the complete right can be named the *Second Quantum Revolution*.

Bibliography

Abbot, D., Davies, P. C. W. and Pati, A. K. (2008). *Quantum Aspects of Life*, Imperial College Press.

Aharonov, Y., Bergmann, P. G. and Lebowitz, J. L. (1964). *Phys. Rev. B* **134**, B1410.

Aharonov Y. and Gruss, E. Y. (2005). Two-time interpretation of quantum mechanics, quant-ph/0507269.

Aharonov Y. and Vaidman, L. (1991). *J. Phys. A* **24**, 2315.

Albert, D. Z. 1992. *Quantum Mechanics and Experience* Harvard Univ. Press, Cambridge, Mass.

Albert, D, and Loewer, B. 1988. *Synthese*, **77**, 195.

Aspect, A, Grangier, P, Roger, G. 1981. *Phys. Rev. Lett.*, **47**, 460.

Aspect, A., Dalibard, J., Roger, G. 1982. *Phys. Rev. Lett.*, **49**, 1804.

Atmanspacher, H. and Primas, H. 2006. *Journal of Consciousness Studies*, **13** (3), 5-50.

Beck, F. and Eccles, J. 2003. Quantum processes in the brain: A scientific basis of consciousness, in *Neural Basis of Consciousness* (Adv. in Consciousness Res., Vol. 49, Ed. N. Osaka), John Benjamins Publ., Philadelphia, PA, p. 141.

Belinskii, A. V. 2003. *Physics-Uspekhi*, **46**, 877.

Bell, J. S. 1964. *Physics*, **1**, 195. Reprinted in [Bell (1987)].

Bell, J. S. 1987. *Speakable and Unspeakable in Quantum Mechanics*, Cambridge Univ. Press, Cambridge.

Bohr, N. 1949. Discussion with Einstein on epistemological problems in atomic physics, in *Albert Einstein: PhilosopherScientist* (The Library of Living Philosophers, Vol. 7, Ed. P. A. Schilpp), Library of Living Philosophers, Evanston, Ill., p. 200. Reprinted in [Wheeler and Zurek eds. (1983)].

Chalmers, D. J. 1996. *The Conscious Mind in Search of a Fundamental Theory* Oxford Univ. Press,New York.

Chernavskii, D. S. 2001. *Sinergetika i Informatsiya: Dinamicheskaya Teoriya Informatsii (Synergetics and Information: Dynamic Information Theory)* Nauka,Moscow.

d'Espagnat, B., 1983. *In Search of Reality* Springer-Verlag, New York.

Deutsch, D. 1997. *The Fabric of Reality: the Science of Parallel Universes and Its Implications*, Allen Lane, New York.

DeWitt, B. S. and Graham, N. (Eds.) 1973. *The Many-Worlds Interpretation of Quantum Mechanics* Princeton Univ. Press, Princeton, NJ.

Donald, M. J. 1990. *Philos. Trans. R. Soc. London*, **A 427**, 43.

Eccles, J. C. 1994. *How the Self Controls its Brain*, Springer-Verlag, Berlin.

Einstein, A., Podolsky, B. and Rosen, N. 1935. *Phys. Rev.*, **47**, 777.

Enz, C. P. 2009. *Of Matter and Spirit*, World Scientific, New Jersey etc.

Everett, H. III, 1957. *Rev. Mod. Phys.* **29** 454. Reprinted in [Wheeler and Zurek eds. (1983)].

Feinberg, E. L. 2004. *Dve Kul'tury. Intuitsiya i Logika v Iskusstve i Kul'ture (Two Cultures. Intuition and Logic in Art and Culture)* in Russian, 3rd ed. Vek-2, Fryazino. [Expanded German edition: Zwei Kulturen. Intuition und Logik in Kunst und Wissenschaften (Berlin: Springer-Verlag, 1998)].

Florensky, P. 1996. *Izbrannye Trudy po Iskusstvu (Selected Works on Art)* in Russian, Izobrazitel'noe Iskusstvo, Moscow.

Gell-Mann, M. and Hartle, J. B. 1990. Quantum mechanics in the light of quantum cosmology, in *Complexity, Entropy, and the Physics of Information* (Santa Fe Institute Studies in the Sci. of Complexity, Vol. 8, Ed. W H Zurek), Addison-Wesley Publ. Co., Redwood City, Calif., p. 425.

Gell-Mann, M. and Hartle, J. B. 1993. *Phys. Rev.*, **D 47**, 3345.

Ginzburg, V. L. 1999. What problems of physics and astrophysics seem now to be especially important and interesting (thirty years later, already on the verge of XXI century)? in Russian, *Physics-Uspekhi*, **42**, 353.

Ginzburg, V. L. 2003. *O Nauke, o Sebe i o Drugikh (About Science, Myself, and Others)* in Russian, 3rd ed. Fizmatlit, Moscow [Translated into English (Bristol: IOP Publ., 2005)].

Giulini, D. et al. 1996. *Decoherence and the Appearance of a Classical World in Quantum Theory* Springer, Berlin.

Grof, S. 1997. *The Cosmic Game*, State University of New York Press, New York.

Hartle, J. 1968. *Am. J. Phys.*, **36**, 704.

Joos, E. and Zeh, H.-D. 1985. *Z. Phys.*, **B 59**, 223.

Josephson, B. D. and Pallikari-Viras, F. 1991. *Found. Phys.*, **21**, 197.

Kapitsa, S. P. 1996. The phenomenological theory of world population growth, *Physics-Uspekhi*, **39**, 57.

Lehner, C. 1997. *Synthese*, **110**, 191.

Lockwood, M. 1989. *Mind, Brain, and the Quantum*, Oxford University Press, Oxford.

Lockwood, M. 1996. Many-minds interpretations of quantum mechanics, *Brit. J. Philos. Sci.*, **47**, 159–188.

Lovelock, J. E. 1990. *Nature*, **344**, 100.

Lossev, A. and Novikov, I. D. 1992. The Jinn of the time machine: nontrivial self-consistent solutions, *Class. Quant. Grav.*, **9**, 2309-2321.

Mansfield, V. 1991. *Possible worlds, quantum mechanics, and Middle Way Buddhism*, in *Symposium on the Foundations of Modern Physics, Joensuu, Finland, 13–17 August 1990* (Eds. P. Lahti, P. Mittelstaedt), World Scientific, Singapore, p. 242.

Markov, M. A. 1991. *O Trekh Interpretatsiyakh Kvantovo— Mekhaniki (On Three Interpretations of Quantum Mechanics)*, in Russian. Nauka,Moscow.

Mensky, M. B. 1979. Quantum restrictions for continuous observation of an oscillator, *Phys. Rev.*, **D 20**, 384-387.

Mensky, M. B. 1979. Quantum restrictions for measurability of motion parameters of a macroscopic oscillator, *Sov. Phys. JETP*, **50**, 667-674.

Menskii, M. B. 2000. Quantum mechanics: New experiments, new applications and new formulations of old questions, *Physics-Uspekhi*, **43**, 585-600.

Mensky, M. B. 2000. *Quantum Measurements and Decoherence. Models and Phenomenology*, Kluwer Acad. Publ., Dordrecht. [Translated into Russian: Fizmatlit, Moscow, 2001]

Menskii, M. B. 2003. Dissipation and decoherence of quantum systems, *Physics-Uspekhi*, **46**, 1163-1182.

Mensky, M. B. 2004. Quantum mechanics, consciousness, and a bridge between the two cultures, *Voprosy Filosofii (Issues in Philosophy)* p.64-74 [in Russian].

Menskii, M. B. 2005. Concept of consciousness in the context of quantum mechanics, *Physics-Uspekhi*, **48** (4), 389-409.

Mensky M. B. 2005. *Human and Quantum World (Weirdness of the quantum world and the miracle of consciousness)* (in Russian), *Vek-2 publishers*, Fryazino.

Menskii M. B. 2007. Quantum measurements, the phenomenon of life, and time arrow: three great problems of physics (in Ginzburg's terminology), *Physics-Uspekhi*, **50**, 397-407.

Mensky M. B. 2007. Reality in quantum mechanics, Extended Everett Concept, and consciousness, *Optics and Spectroscopy*, **103**, 461-467.

Mensky M. B. 2007. Postcorrection and mathematical model of life in Extended Everett's Concept, *NeuroQuantology*, **5**, 363-376, (www.neuroquantology.com, arxiv:physics.gen-ph/0712.3609).

Mensky M. B. von Borzeszkowski, H. 1995. Position measurement for a relativistic particle: Restricted-Path-Integral analysis, *Phys. Lett.*, **A 208**, 269-275.

Meyenn, K. von, (ed.). 1996. *Wolfgang Pauli. Wissenschaftlicher Briefwechsel*, Band IV, Teil I: 1950-1952. Springer, Berlin.

Panov, A. D. 2001. *Physics-Uspekhi*, **44**, 427.

Panov, A. D. 2008. *Universal Evolution and the Problem of Search for Extra Terrestrial Intelligence (SETI)*, in Russian, *LKI editors*, Moscow.

Panov, A. D. On the methodological problems of cosmology and quantum gravity (in Russian), to be published.

Paz, J. P. and Zurek, W. H. 1993. *Phys. Rev.*, **D 48**, 2728.

Penrose, R. 1991. *The Emperor's New Mind: Concepting Computers, Minds, and the Laws of Physics* Penguin Books, New York.

Penrose, R. 1994. *Shadows of the Mind: a Search for the Missing Science of Consciousness* Oxford Univ. Press, Oxford.

Penrose, R. 2004. *The Road to Reality*, Jonathan Cape, London.

Popov, M. A. 2003. *Physics-Uspekhi*, **46**, 1307.

Rubin, M. A. 2003. *Found. Phys.*, **33**, 379.

Satprem. 1970. *Sri Aurobindo ou L'aventure de la conscience*, Bucher Chastel.

Saunders, S. 1993. *Phys. Lett.*, **A 184**, 1.

Schrödinger, E. 1944. *What is Life? The Physical Aspect of the Living Cell*, The Cambridge Univ. Press, Cambridge.

Schrödinger, E. 1958. *Mind and Matter*, Cambridge University Press, Cambridge.

Smolin,   L.   2009.   The   unique   universe.
http://physicsworld.com/cws/article/indepth/39306

Squires, E. 1994. *The Mystery of the Quantum World*, 2nd ed. IOP Publ., Bristol.

Stapp, H. P. 2001. *Found. Phys.*, **31**, 1465.

Tegmark, M. 1998. *Fortschr. Phys.*, **46**, 855.

Teilhard de Chardin, P. 1959. *The Phenomenon of Man*, Harper and Row, New York.

Vaidman, L. 2002. *The many-worlds interpretation of quantum mechanics*, in *Stanford Encyclopedia of Philosophy* (Electronic Resource) (Ed. E N Zalta) (Stanford, Calif.: Metaphysics Research Lab, Centre for the Study of Language and Information, Stanford Univ., 2002); http://plato.stanford.edu/archives/sum2002/entries/qm-manyworlds/

Valiev, K. A. 2005. *Physics-Uspekhi*, **48**, 1.

Villars, C. N. 1983. *Psychoenergetics*, **5**, 1.

Von Neumann, J. 1932. *Mathematische Grundlagen der Quantenmechanik* (Berlin: J. Springer, 1932) [Translated into English: *Mathematical Foundations of Quantum Mechanics*, Princeton Univ. Press, Princeton, NJ, 1955]

Wallace, . A. 2007. *Hidden Dimensions: The Unification of Physics and Consciousness*, Columbia University Press, New York.

Wheeler, J. A., and Zurek, W. H., editors. 1983. *Quantum Theory and Measurement*. Princeton University Press, Princeton.

Whitaker, A. 2000. *Many minds and single mind interpretations of quantum theory*, in *Decoherence: Theoretical, Experimental, and Conceptual Problems: Proc. of a Workshop, Bielefeld, Germany, November 1998* (Eds P. Blanchard et al.), Springer, Berlin, 2000, p. 299.

Wigner, E. P. 1961. Remarks on the mind-body question, in L. G. Good, editor, *The Scientist Speculates*, Heinemann, London, pages 284–302. Reprinted in [Wheeler and Zurek eds. (1983)].

Zeh, H.-D. 1970. *Found. Phys.*, **1**, 69.

Zeh, H.-D. 1992. *The Physical Basis of the Direction of Time*, 2nd ed. Springer-Verlag, Berlin.

Zeh, H.-D. 2000. The Problem of Conscious Observation in Quantum Mechanical Description, *Found. Phys. Lett.*, **13**, 221-233.

Zurek, W. H. 1981. *Phys. Rev.*, **D 24**, 1516.

Zurek, W. H. 1982. *Phys. Rev.*, **D 26**, 1862.

Zurek, W. H. 1998. *Philos. Trans. R. Soc. London*, **A 356**, 1793.

Index